WITHDRAWN

Medical and Biologic Effects of Environmental Pollutants

MANGANESE

*Committee on
Biologic Effects of
Atmospheric Pollutants*

DIVISION OF MEDICAL SCIENCES
NATIONAL RESEARCH COUNCIL

NATIONAL ACADEMY
OF SCIENCES
WASHINGTON, D.C.   1973

Other volumes in the Medical and Biologic Effects of Environmental Pollutants series (formerly named Biologic Effects of Atmospheric Pollutants):

ASBESTOS: The Need for and Feasibility of Air Pollution Controls
(ISBN 0-309-01927-3)
FLUORIDES (ISBN 0-309-01922-2)
LEAD: Airborne Lead in Perspective (ISBN 0-309-01941-9)
PARTICULATE POLYCYCLIC ORGANIC MATTER (ISBN 0-309-02027-1)

NOTICE: The project that is the subject of this report was approved by the Governing Board of the National Research Council, acting in behalf of the National Academy of Sciences. Such approval reflects the Board's judgment that the project is of national importance and appropriate with respect to both the purposes and the resources of the National Research Council.

The members of the committee selected to undertake this project and prepare this report were chosen for recognized scholarly competence and with due consideration for the balance of disciplines appropriate to the project. Responsibility for the detailed aspects of this report rests with that committee.

Each report issuing from a study committee of the National Research Council is reviewed by an independent group of qualified persons according to procedures established and monitored by the Report Review Committee of the National Academy of Sciences. Distribution of the report is approved, by the President of the Academy, upon satisfactory completion of the review process.

The work on which this publication is based was performed pursuant to Contract No. 68-02-0542 with the Environmental Protection Agency.

Available from

Printing and Publishing Office, National Academy of Sciences
2101 Constitution Avenue, N.W., Washington, D.C. 20418

LIBRARY OF CONGRESS CATALOGING IN PUBLICATION DATA

National Research Council. Committee on Medical and Biologic Effects of Environmental Pollutants.
Manganese.

(Its Biologic effects of environmental pollutants) Bibliography: p.
1. Manganese–Environmental aspects. 2. Manganese–Physiological effect. 3. Air–Pollution. I. Title.
QH545.M3N37   1973   614.7'12   73-18174
ISBN 0-309-02143-X

Printed in the United States of America

PANEL ON MANGANESE

JAN LIEBEN, American Viscose Division, FMC Corporation, Philadelphia, Pennsylvania, *Chairman*
GILBERT L. DeHUFF, JR., Division of Ferrous Metals, U. S. Bureau of Mines, Arlington, Virginia
K. MICHAEL HAMBIDGE, Department of Pediatrics, University of Colorado Medical Center, Denver, Colorado
JAMES W. LASSITER, SR., Department of Animal Science, University of Georgia, Athens, Georgia
PAUL M. NEWBERNE, Department of Nutritional Pathology, Massachusetts Institute of Technology, Cambridge, Massachusetts
ARTHUR J. RIOPELLE, Department of Psychology, Louisiana State University, Baton Rouge, Louisiana

CHARLES D. FOY, Plant Industry Station, U. S. Department of Agriculture, Beltsville, Maryland, *Consultant*
N. E. WHITMAN, American Industrial Hygiene Association, Allentown, Pennsylvania, *Consultant*

JAROSLAV J. VOSTAL, Department of Pharmacology and Toxicology, University of Rochester Medical Center, Rochester, New York, *Associate Editor*
BERNARD WEISS, Department of Radiation Biology and Biophysics, University of Rochester Medical Center, Rochester, New York, *BEAP Committee Liaison Representative*

JOHN REDMOND, JR., Division of Medical Sciences, National Research Council, Washington, D.C., *Staff Officer*

COMMITTEE ON BIOLOGIC EFFECTS OF ATMOSPHERIC POLLUTANTS

HERSCHEL E. GRIFFIN, Graduate School of Public Health, University of Pittsburgh, Pittsburgh, Pennsylvania, *Chairman*
DAVID M. ANDERSON, Environmental Quality Control Division, Bethlehem Steel Corporation, Bethlehem, Pennsylvania
VINTON W. BACON, Department of Civil Engineering, College of Applied Science and Engineering, University of Wisconsin, Milwaukee, Wisconsin
ANNA M. BAETJER, Department of Environmental Medicine, School of Hygiene and Public Health, The Johns Hopkins University, Baltimore, Maryland
RICHARD U. BYERRUM, College of Natural Science, Michigan State University, East Lansing, Michigan
W. CLARK COOPER, 2180 Milvia Street, Berkeley, California
ROBERT I. HENKIN, National Heart and Lung Institute, National Institutes of Health, Bethesda, Maryland
SAMUEL P. HICKS, Department of Pathology, University of Michigan Medical Center, Ann Arbor, Michigan
IAN T. T. HIGGINS, School of Public Health, University of Michigan, Ann Arbor, Michigan
JAN LIEBEN, American Viscose Division, FMC Corporation, Philadelphia, Pennsylvania
JAMES N. PITTS, JR., Statewide Air Pollution Control Center, University of California, Riverside, California
I. HERBERT SCHEINBERG, Albert Einstein College of Medicine, Bronx, New York
RALPH G. SMITH, Department of Environmental and Industrial Health, School of Public Health, University of Michigan, Ann Arbor, Michigan
GORDON J. STOPPS, Environmental Health Branch, Ontario Department of Health, Toronto, Ontario, Canada
F. WILLIAM SUNDERMAN, JR., Department of Laboratory Medicine, University of Connecticut School of Medicine, Farmington, Connecticut
BENJAMIN L. VAN DUUREN, Institute of Environmental Medicine, New York University Medical Center, New York, New York
JAROSLAV J. VOSTAL, Department of Pharmacology and Toxicology, University of Rochester Medical Center, Rochester, New York
BERNARD WEISS, Department of Radiation Biology and Biophysics, University of Rochester Medical Center, Rochester, New York

T. D. BOAZ, JR., Division of Medical Sciences, National Research Council, Washington, D.C., *Executive Director*

Acknowledgments

This document was prepared by the Panel on Manganese with the assistance of the Committee on Biologic Effects of Atmospheric Pollutants* of the National Research Council.

The total document has been reviewed by the entire Panel under the direction of the Panel chairman, Dr. Jan Lieben. The author of the chapter on manganese in the ecosystem was Mr. Gilbert L. DeHuff, Jr. The chapter on manganese and plants was drafted by Dr. Charles D. Foy. Dr. Paul M. Newberne drafted the chapter on input and disposition of managanese in man. The biochemistry and metabolic role of manganese were described by Drs. James W. Lassiter, Sr., and K. Michael Hambidge. The interrelations of manganese and catecholamines were discussed by Drs. K. Michael Hambidge and James W. Lassiter, Sr. Material on the epidemiology of manganese intoxication and permissible air concentrations was prepared by Dr. Jan Lieben and that on neurobehavioral effects of manganese deficiency and toxicity and on methyl manganese tricarbonyl compounds, by Dr. Arthur J. Riopelle. The appendix on environmental sampling and analysis of manganese was prepared by Mr. N. E. Whitman.

*The title of the Committee has since been changed to Committee on Medical and Biologic Effects of Environmental Pollutants, but its composition and function remain the same.

The preparation of the document was assisted by the comments of the members of the Committee on Biologic Effects of Atmospheric Pollutants—particularly, Dr. Jaroslav J. Vostal, who served as associate editor, and Dr. Bernard Weiss, who served as Committee liaison. The Panel is indebted to the four anonymous reviewers, selected by the associate editor, who offered many useful comments on the original manuscript. The entire report was edited by Mr. Norman Grossblatt, Editor for the Division of Medical Sciences.

Dr. Robert J. M. Horton of the Environmental Protection Agency (EPA) was of invaluable assistance in obtaining documents and translations and in offering counsel. Dr. Douglas I. Hammer was the EPA liaison officer. Informational assistance was provided by the National Research Council Advisory Center on Toxicology, the National Academy of Sciences Library, the National Library of Medicine, the National Agricultural Library, the Library of Congress, and the Air Pollution Technical Information Center.

Acknowledgment of the assistance given by the Environmental Studies Board of the National Academy of Sciences and National Academy of Engineering and by various divisions of the National Research Council is hereby recorded.

CHARLES L. DUNHAM
Chairman, Division of Medical Sciences

Contents

1	Introduction	1
2	Manganese in the Ecosystem	3
3	Manganese and Plants	51
4	Input and Disposition of Manganese in Man	77
5	Biochemistry and Metabolic Role of Manganese	83
6	Manganese Toxicity and Catecholamines	92
7	Epidemiology of Manganese Intoxication	101
8	Permissible Air Concentrations of Manganese	114
9	Neurobehavioral Effects of Manganese Deficiency and Toxicity	116
10	Manganese Tricarbonyl Compounds	126
11	Conclusions	132
12	Recommendations for Future Research	137
	Appendix: Environmental Sampling and Analysis of Manganese	141
	References	149
	Index	181

1

Introduction

The Environmental Protection Agency (EPA) is required by the Clean Air Act, as amended, to promulgate regulations for the control of air pollutants that are deleterious to human health and welfare. To meet the requirement, the Agency must know the effects of those pollutants. This knowledge permits the Administrator of EPA to decide whether regulation is necessary and, if so, which of several control strategies is appropriate.

This document is one of a series that the EPA has asked the National Academy of Sciences to prepare. Although the main thrust of the document is to examine the effects of airborne manganese, it also considers biologic intake from all other routes.

Manganese is thought by some to have been named after the Latin "magnes" (magnet) for the supposed magnetic properties of pyrolusite, its principal ore; others think it was named after the Italian word for "magnesia." It was first recognized as an element by Scheele and others and was first isolated by Gahn in 1774.

Manganese has been used in making steel since 1839. But it did not come into general use until the later part of the nineteenth century, after Robert Mushet's discovery that addition of manganese overcame problems in making steel with the Bessemer process.[129]

The first report of human illness associated with excessive exposure to manganese was in 1837.[118] Around the beginning of the twentieth century, the hazard of manganism was "rediscovered." As better analytic techniques have been developed and more has been discovered about the complex chemistry of organisms, it has become apparent that man and many lower animals require some manganese if they are to function properly but that excessive manganese can be toxic.

This document presents a critical evaluation of the literature, through July 1, 1972, on manganese and its biologic effects.

2

Manganese in the Ecosystem

NATURAL OCCURRENCE OF MANGANESE

Distribution in the Earth's Crust

Manganese is widely distributed in the earth's igneous, sedimentary, and metamorphic rocks, and it is the principal metallic constituent of nodules and other surficial deposits that cover large areas of the deep ocean floors. It has been calculated to be the twelfth most abundant element in the earth's crust, assuming the crust to have a depth of about 16 km. Of the more commonly recognized metals, manganese is exceeded in abundance by aluminum, iron, magnesium, and titanium but is much more abundant than nickel, vanadium, copper, uranium, zinc, and lead. Although the calculations are somewhat arbitrary, manganese can be said to constitute about 0.10% (1,000 ppm) of the earth's total crust.[98, 202, 307, 329] The crust contains about 50 times as much iron, one-fifth as much nickel, and one-tenth as much copper.[98]

Manganese is next to iron in the atomic series, is similar to it in chemical behavior, and is often closely associated with it in its natural occurrence. Unlike iron, it is not ferromagnetic; however, some of its alloys and compounds are.[468] In its compounds, manganese can have valences of 1, 2, 3, 4, 6, and 7; it is divalent in the most stable salts, and the dioxide (MnO_2) is the most stable oxide.[327] Compounds are now also known, including organometallic complexes, in which manganese has valences of -3, -1, zero, and 5.[219]

In presenting a table of the distribution of the elements in the earth's crust, Turekian and Wedepohl state that

> any compilation is necessarily subject to great uncertainties in the reliability of the analytical work, the sampling, and the interpretations, both of the original investigator and the compiler. Hence the accompanying table should be accepted not so much as a doctrine but as a motion on the floor to be debated, and amended or rejected.[497]

This premise holds for much of the following quantitative information on manganese in the ecosystem; such data are of value chiefly in presenting orders of magnitude that can be considered reasonable in most instances. Where substantial differences occur among the data from recognized sources, it seems advisable to present a fair sampling of the diverse material.

Turekian and Wedepohl suggest the following manganese contents, expressed in parts per million, for three major components of the earth's crust:[497]

igneous rock: ultrabasic, 1,620; basaltic, 1,500; high-calcium granitic, 540; low-calcium granitic, 390; and syenites, 850

sedimentary rock: shales, 850; sandstones, an order-of-magnitude estimate of zero; and carbonates, 1,100

deep-sea sediment: carbonate, 1,000; and clay, 6,700

At least 100 minerals—including sulfides, anhydrous and hydrous oxides, carbonates, anhydrous and hydrous silicates, anhydrous and hydrous phosphates, arsenates, tungstates, and borates—contain manganese as an essential element; and it is an accessory element in perhaps more than 200 others.[227] Oxides are the most important ore minerals of manganese. They frequently grade in composition from one to another, giving rise to considerable confusion in terminology and specific identification, unless recourse is made to X ray or other sophisticated laboratory methods of analysis; even then, specific identification may be difficult. Physically, the manganese oxide minerals grade from quite hard and dense to friable and earthy. Many are of secondary origin, formed under the influence of weathering. Pyrolusite, a mineral form of manganese dioxide, is one of the more common manganese oxide minerals. The principal manganese carbonate ($MnCO_3$) ore mineral is rhodochrosite, which is a member of an isomorphous series with the carbonates

Manganese in the Ecosystem

of iron, calcium, magnesium, zinc, and cobalt. The only silicate ore mineral has been the oxysilicate, braunite. However, the most common manganese silicate ($MnSiO_3$) mineral is rhodonite, as yet too refractory to have served as an ore.[129] Manganese does not occur naturally as the metal.

Manganese ore deposits are widespread throughout the tropical, subtropical, and warmer temperate zones of the earth. The largest deposits are found at Chiatura in the Caucasus and at Nikopol in the Dnieper Basin (both in the USSR), in mainland China, in Brazil, in India, in Australia, and in Africa in the Republic of South Africa, Gabon, Zaire (Congo-Kinshasa), Ghana, and Morocco. The average grade of the ores or concentrates shipped from these deposits varies from 40% or less manganese for the Dnieper Basin ores to 50% or more for some of the African ores. Deposits of lower-grade manganese ores and protores occur in these countries and in countries of more temperate latitudes, including the United States, where weathering has not been sufficient for extensive enrichment.

In the United States, manganese deposits have been well distributed through the southern Appalachian and Piedmont regions, the Batesville district of Arkansas, and many of the western states. These deposits have been exhausted in terms of mining at a profit at existing and appreciably higher prices. In most instances, however, the surrounding areas can be assumed to have an abnormally high concentration of manganese in the soil and possibly in the water. Apart from these concentrations, there are large low-grade manganese deposits extending for miles along both sides of the Missouri River in South Dakota and large low-grade deposits in the Cuyuna Range of Minnesota, in the Artillery Mountains region of northwestern Arizona, in the Batesville district of Arkansas, in Aroostook County in Maine, and to a lesser extent in the Gaffney-Kings Mountain district of North Carolina and South Carolina. Because of their large tonnages, averaging from 4 to 10 or 15% manganese, these deposits constitute resources for possible exploitation.[127,128]

Soil

After studying the trace-element content of soils in Canada, Great Britain, and two small areas in Wisconsin, Warren et al.[530] suggest an average manganese content of 800 ppm, with deviation of 50% in either direction from the average, as a normal content for soils. They qualify their paper "merely as a report of a 'reconnaissance in force.' The resources at our disposal made it impossible for us to collect and analyze sufficient samples to enable statistically valid conclusions." For comparative purposes, they show an average of 850 ppm developed from Vinogradov's many recorded analyses.[517] Swaine[474] gives a range of 200–3,000 ppm for "total content of Mn in most soils." Vinogradov observes that great enrichment in manganese is to be found in the soils on or near manganese deposits and gives 6.12% manganese content for the soils at Chiatura, the locale of the Caucasian ores. His highest concentrations are 10–15%, for the acid soils of Hawaii; and his lowest, 0.000% and 0.00%, for some samples from Sweden and China, respectively.[517]

Analysis of samples taken by Wright et al.,[551] in the course of studying virgin profiles of four great Canadian soil groups developed on glacial till, gave a manganese content of 250–1,380 ppm. Bowen[69]—digesting data from Swaine,[474] Vinogradov,[517] Bear,[45] and others—gives 850 ppm as the mean manganese content and 100–4,000 ppm as the range for oven-dried soils. In computing the range, he omitted soils from the vicinity of mineral deposits. In tabulating trace-element analyses obtained from eight representative Scottish soil profiles out of more than 100 studied, Swaine and Mitchell[475] show a range for total manganese of 50–7,000 ppm in air-dried soil. Their study shows a range for extractable manganese (i.e., the manganese available to plants) of from less than 0.8 to 130 ppm (soluble in 2.5% acetic acid) in air-dried soils. They stress the importance of the geologic parent material in determining the status of the soil with respect to trace elements.

In a program that began in 1961, the U.S. Geological Survey collected soil samples from a depth of approximately 20 cm at 863 sites throughout the conterminous United States. The results of analysis showed a range of manganese content of from less than 1 to 7,000 ppm and an arithmetic mean of 560 ppm.[446]

To obtain background data for a recent study of the possibility of metallic pollution of soils, Cannon and Anderson collected 39 soil samples from several geologic settings at soil depths of about 5–20 cm in remote areas of the United States judged to be relatively free of contamination. The mean of the analyses of these samples for manganese was 660 ppm, and the median was 500 ppm.[85]

Ocean-Floor Deposits and Seawater

The manganese deposits of the deep ocean floors have been known for many years but have become of much interest only recently. They are usually in the form of manganese oxide nodules with concentric growth, often around a shark's tooth, grain of sand, lump of clay, or other barren nucleus. They are found over large areas of the Pacific, Atlantic, and Indian Ocean floors beyond the continental shelves. The nodules contain cobalt, nickel, and copper in sufficient quantities to make the deposits probably of more interest for these elements than for manganese. However, their concentration are such that some areas favor cobalt, some nickel, and some copper. Iron also is commonly an associate element. There is a unique deposit on the Blake Plateau of the Atlantic off the Georgia and Florida coasts to depths of about 1,068 m. It consists of a continuous solid pavement of manganese oxide on which a wheeled undersea vessel has driven where the surface has been swept clean by the Gulf Stream. Elsewhere, the pavement is covered with manganese oxide nodules; and at still other locations, there are nodules on the ocean floor without a pavement.

Although manganese analyses of up to 50% have been reported for these deposits—"almost 80% MnO_2" on a dry-weight basis[349 (p.227)]—the average for possible minable areas appears to be considerably less, probably around 20%[200] or 25% manganese. For a group of 54 samples from the Pacific Ocean, Mero[349 (p.180)] reports a maximal manganese content of 41.1% (dry-weight basis as determined by x-ray emission spectrography), an average of 24.2%, and a minimum of 8.2%, and for four samples from the Atlantic Ocean, a maximum of 21.5%, an average of 16.3%, and a minimum of 12.0%.

When considering the concentration of the less-abundant elements in seawater, it cannot be assumed that the sea is a completely homogeneous medium. "The surface waters are generally much less characteristic of the hydrosphere than are samples taken from a depth of 5,000 feet or so."[349 (p.24)] Biologic activity and local conditions are factors responsible for variations.

The concentration of manganese dissolved in seawater has been given by Goldberg[200] as 0.002 mg/liter (0.002 ppm), the same as reported for nickel and vanadium (each) and one fifth the concentration reported for iron, zinc, and aluminum (each);[349 (p.26)] by Sverdrup et al.[473] as 0.001–0.01 g/ton (about 0.001–0.01 ppm), with chlorinity equal to 19 ppt (parts per thousand); and by Turekian[496] (with Slowey[457] as the source) as 0.4 ppb or 0.4 µg/liter (0.0004 ppm), using neutron-activation analysis with the salinity of seawater taken at 35 ppt. Concentrations differed among the various sources with respect to the other elements that have been presented above for comparative purposes. Furthermore, some data are qualified, in that "many of the values given were obtained from a single set of analyses of a sample of surface water."[349 (p.24)] Goldberg and Arrhenius[201 (p.196)] determined the distribution between particulate (>0.5 µm) and dissolved manganese at different depths at one station in the Pacific, using chemical analysis, as shown in Table 2-1. They reported that experiments with a filter of

TABLE 2-1 Distribution of Particulate and Dissolved Manganese, Pacific Ocean, lat. 26°28.5′N, long. 170°23.0′W (Station 129)[a]

Depth, m	Dissolved Manganese, µg/liter	Particulate Manganese, µg/liter	Particulate Fraction of Total, %
0	2.8	0.22	7.3
200	4.7	0.08	1.7
1,000	5.0	0.06	1.2
1,500	2.0	0.05	2.4
2,000	2.1	0.09	4.1
2,500	1.2	0.02	1.6
3,000	1.7	0.12	6.6
3,500	0.4	0.07	15

[a] Derived from Goldberg and Arrhenius.[201]

5- to 10-nm pore size obtained approximately the same quantity of retained manganese, "suggesting that 85 percent or more of the element prevails in true solution." It is to be noted that their concentrations for manganese in solution at this station ranged from 0.4 to 5.0 µg/liter (0.0004 to 0.005 ppm), with the higher values in the upper half of the section sampled.

Fresh Water

Skougstad,[456] concerned with the determination of minor-element content of fresh waters, details the problems involved in the sampling and analysis of water. He points out that "water in nature is almost always a continuously changing material, so that a sample collected at one moment may have a different composition from one collected at the same place sometime later." This is, of course, even more the case for stream waters that receive waste discharges. He settled on emission spectroscopy as the method of analysis most suitable for most of the minor elements of interest, including manganese, taking into account the sensitivity of the method and the important factors of time and cost.

Livingstone,[310] in compiling representative chemical data for the lake and river waters of the world, discusses the complexities of these waters and the factors that influence their composition, comments on the frequent lack of discrimination between dissolved and colloidal or suspended material in the data, and discusses the problems of sampling and analysis. He comments that very little is known about the state of manganese in these waters. Pronounced changes in manganese concentration occur with changes in depth in stratified lakes.

The most common situation appears to be one in which the manganese content is high in the reduced bottom water; it reaches high concentrations at a somewhat shallower depth than iron, presumably because manganous ion is released from the bottom at a slightly higher redox potential than ferrous iron (Hutchinson,[251] 1957, p. 809)....

Ohle (1934)[382] studying lakes in North Germany found a total manganese content between less than 5 and as much as 200 ppb. The mean was 25 ppb. One lake, Trammersee, had a variation in manganese throughout a single year that covered almost the entire range, from less than 5 ppb to 133 ppb. Juday, Birge, and Meloche (1938)[271] found comparable amounts, 3 to 23 ppb in the surface waters of 8 Wisconsin lakes. The deep water of one lake contained 1200 ppb. Uniformly high manganese contents have been recorded for some waters—for example, 50 to 250 (mean of 140 ppb) for Linsley Pond (Hutchinson, 1957, p. 803-804)[251] and 80 to 120 ppb for the Mississippi River at Fairport, Iowa (Wiebe, 1930).[539] The mean for the rivers of the USSR is 11.9 ppb (Konovalov, 1959),[288] but the global average is probably somewhat higher....

Lohammar (1938)[314] has provided a very substantial body of information on the manganese content of waters of Sweden. There seems to be a slight difference in the waters of northern and southern Sweden in this respect. In north Sweden the range was >10-60 ppb, with a mean of 33 ppb, and in south Sweden >10-850 ppb, with a mean of 44 ppb. Waters from northern Sweden have a much higher iron content than those from southern Sweden, and there seems very little doubt that the Fe/Mn ratio is significantly higher for the northern (30) than for the southern (5) waters.

In investigating waters from 440 Maine lakes, Kleinkopf,[285,286] apparently sampling mostly surface waters, found the manganese content to range from 0.02 to 87.5 ppb, with a mean of 3.8 ppb.[310]

In 1966, Konovalov et al.[289] reported the manganese in solution and in suspended material in various river waters of the European USSR, as shown in Table 2-2. In the suspended material, there was considerably more manganese than any of the 10 other elements determined, except iron. Manganese contents were determined by chemical analysis of two to six samples collected in periods of normal and high water, usually from close to the delta. From the total quantity of suspended material in the water, the authors computed the quantity of suspended manganese per liter. They note that, in addition to material that enters the river from drainage of the watershed area, the suspended substances include material precipitated in the river. Also, there is a continual exchange of manganese into solution from the colloids of the suspended material, and an increase of manganese in solution was observed with increased mate-

TABLE 2-2 Manganese Content in Suspended Substances in Water[a]

River	Site	Date of Sample Collection, month/day/year	Concentration of Suspended Substances, g/liter	Manganese Concentration in Suspension % of Suspended Material	μg/liter	Manganese Concentration in Water, μg/liter
		Barintsev and White Seas Basin				
North Dvina	Ust' Pirega	5/13/1955	0.101	b	b	b
Pechora	Ust' Tsil'ma	6/19/1954	0.038	0.063	24	13.2
	Ust' Tsil'ma	10/17/1954	0.016	0.058	9	8.0
		Baltic Basin				
Neman	Sovetsk	4/15/1954	0.010	b	b	13.8
	Sovetsk	10/8/1954	0.043	b	b	1.0
Western Dvina	Plyavinas	4/12/1954	0.073	0.246	180	19.3
Narva	Narva	5/12/1954	0.003	b	b	14.0
		Black and Azov Seas Basin				
Dnestr	Bendery	4/3/1954	0.397	0.136	540	25.3
	Bendery	5/25/1955	0.088	0.225	196	30.0
	Bendery	8/7/1955	0.241	0.123	298	8.4
Dnepr	Berislav	5/28/1955	0.120	0.242	290	14.3
Kuban'	Slavyanskaya	4/19/1954	0.245	0.102	250	3.8
	Slavyanskaya	9/10/1954	0.156	0.175	272	2.5
	Slavyanskaya	7/25/1955	0.242	0.166	405	8.0
	Sl yanskaya	10/25/1955	0.062	0.079	49	4.0
Rioni	Poti	5/26/1954	0.770	1,160	8,900	40
	Poti	9/14/1954	0.506	0.680	3,440	96
	Poti	5/31/1955	0.369	0.580	2,140	52
	Poti	9/25/1955	0.478	5,900	28,200	124
Dunay	Kiliya	4/7/1954	3.11	0.100	3,110	70
	Kiliya	9/30/1954	0.057	0.274	156	10
		Drainageless Basin of Caspian Sea				
Kura	Sal'yany	5/17/1954	3.23	0.360	11,600	160
	Sal'yany	8/31/1954	0.288	0.125	359	4.6
Terek	Kargalinskaya	4/26/1954	1.90	0.129	2,450	23.2
	Kargalinskaya	10/25/1954	0.343	0.129	442	12.0
	Kargalinskaya	8/8/1955	2.06	0.132	2,720	6.8
	Kargalinskaya	12/9/1955	0.234	0.372	870	8.2
Ural	Gur'ev	5/14/1954	0.698	0.174	1,100	3.2
	Gur'ev	11/11/1954	0.023	b	b	3.6
Volga	Volgograd	2/25/1955	0.103	0.122	126	3.0
	Volgograd	6/6/1955	0.033	0.252	83	5.2
	Volgograd	8/29/56	0.030	0.225	67.5	37.5

[a] Derived from Konovalov et al.[289]
[b] Not determined.

rial in suspension, particularly in periods of high water. The manganese in solution ranged from 1.0 to 160 μg/liter; the manganese content of suspended material, from 0.058% to 5.9%; and the calculated suspended manganese, from 9 to 28,200 μg/liter. The Rioni River, which is fed by tributaries that flow through the Caucasian manganese deposits, yielded consistently high values.

Durum and Haffty[145,146] have made spectrographic analyses for minor dissolved elements in waters from 15 large rivers of the United States and Canada. Of 52 determinations for manganese, the highest values were 185 μg/liter (for one of three samples from the Mississippi River near Baton Rouge, Louisiana, the other two yielding values of 12 and 46 μg/liter) and 181 μg/liter (for a sample from the Yukon River at Mountain Village, Alaska; no other determinations were made here). The median for all 52 samples was 20 μg/liter (20 ppb); the range was 0–185 μg/liter.

Other Occurrences

Of 27 samples of mineral waters collected before 1920 from wells, springs, geysers, and mine waters, an appreciable number were reported to have a trace of manganese.[97] It is probable, however, that these reports of trace actually reflect nondetermination for or nondetection of manganese or inadequacies of analytic methods in use at the time for quantifying traces discovered. Even with present methods, Kroner and Kopp[292] have pointed out that, in working with water with low concentrations of trace elements, "failure to detect a particular element does not mean that the element is absent"; they cite adsorption on suspended matter or container walls as examples. Hewett[226 (p.563)] placed the maximal manganese content of spring waters at about 117 ppm.

From experimental work, it was observed more than 40 years ago that several genera of bacteria common to soils and oceanic muds precipitated manganese oxides from manganese salts.[226] For many years, bacteria have been considered to be an agent in the formation of bog manganese ores. Recent research has demonstrated that manganese can be separated out of refractory manganiferous materials by the action of by-products formed as a result of microbial metabolism. For a number of years, the U.S. Bureau of Mines conducted research on the use of bacteria to leach manganese from low-grade and refractory manganiferous materials. The extractions in the laboratory ranged from 71.7% to 99.9%, with appreciable concentration of the manganese in the final products.[399]

Insofar as is known, all plants and animals contain manganese,[226,309,442] although the quantities may be minute.

MAN-MADE SOURCES OF MANGANESE

In depicting the natural occurrences of manganese, it was at times difficult to draw a line between occurrences unaffected by man's activities and those so affected, which were intended to be discussed in following sections. We have attempted to separate the natural from the man-made sources on a basis of degree of significance balanced to some extent by considerations of context. As a result, some material in each section might appear to belong in the other.

Patterns of Consumption

Manganese is essential to the tonnage production of steel and is important in the production of aluminum, magnesium, and cast irons. It is a necessary constituent of manganese bronzes used for ship propellers and of some other copper alloys. Some manganese ores are used as essential ingredients of the common dry-cell battery, and these and others have important chemical uses.

Manganese ores are mined by both open-pit and underground methods; much of the large-tonnage production comes from open pits. Various degrees of mechanization are to be found, both above and below ground. Large bucket-wheel excavators in conjunction with long conveyors, in use in the Dnieper Basin to remove both overburden and ore, are an example of some of the more advanced open-pit mechanization.

Very little, if any, manganese ore is now shipped without some form of beneficiation, although it may be only simple washing and screening. Crushing, heavy-media separation, jigging, tabling, flotation, and magnetic separation are practiced. Most of these are wet operations, but some pneumatic concentration has been used in Morocco and possibly elsewhere. Roasting, sintering, and nodulizing operations are or have been used for some ores.

World production by countries for 1968–1970 is shown in Table 2-3.

Most of the concentrate goes into the production of ferromanganese, which is used primarily in the production of steel. Roughly 90% or more of the manganese consumed in the United States is used in the production of iron and steel, in which its principal purpose is to nullify the harmful effects of sulfur. No satisfactory substitute has been found for it in this function,[42] and it serves as a deoxidizer at the same time. It is also added to steel as an alloying agent to provide strength, toughness, hardness, and hardenability. Standard high-carbon ferromanganese—with a manganese content generally of 78–82% (formerly 74–82%), a carbon content of approximately 7%, and the balance largely iron—is made in either blast or electric furnaces. Silicomanganese (in effect, a high-silicon ferromanganese)—usually with a manganese content of 65–68%, a silicon content of 18–20%, and the balance chiefly iron—and the refined grades of ferromanganese (medium carbon, low carbon, and special grades of high carbon) are made in electric furnaces. One company uses fused-salt electrolysis to produce low-carbon ferromanganese. Spiegeleisen—with a manganese content of approximately 20% and the balance largely iron— is used almost entirely for cast irons and steel and is made in either blast or electric furnaces.

Electrolytic manganese metal—99.9% manganese (metallic basis)—is used in the production of steel alloys, aluminum alloys, and copper alloys and elsewhere when it is desired to add manganese without introducing impurities. Its use in the production of ordinary carbon steel has resulted in some operating advantages. Manganese gives strength, hardness, and stiffness to aluminum, and it is usually added in the form of a manganese–aluminum master alloy, or a manganese–aluminum briquet, made from the electrolytic metal. Electrolytic manganese metal is usually made from metallurgic-grade ores or from high-manganese slag obtained from the electric-furnace production of ferromanganese. For a number of years, its production in the Republic of South Africa was solely from effluent solutions obtained in the processing of uranium ores. Its high cost, approximately three times that of ferromanganese, limits its use. Production in the United States in 1970 was 29,000 tons, compared with 835,000 tons of ferromanganese and 193,000 tons of silicomanganese.[130]

Overseas, manganese metal of lesser purity is made in electric furnaces and by the older Goldschmidt process of aluminothermic reduction.

A large quantity of manganese enters the steel-making process as a component of iron ores and of recycled open-hearth slags fed to the pig-iron blast furnaces. The total of this manganese and that contained in iron and steel scrap fed to the steel furnaces has, in the past, been calculated to exceed that introduced as ferroalloy.[127,129] It probably still does, even with the technologic changes in the production of steel that have since taken place.

The common dry-cell battery uses manganese dioxide as the depolarizer in the cell. Either battery-grade dioxide ore, synthetic dioxide, or a mixture of the two is used. Synthetic manganese dioxide is made either

TABLE 2-3 World Production of Manganese Ore[a]

Country	% Manganese[b]	Ore Production, tons		
		1968	1969	1970[c]
North America				
Mexico[d]	35+	65,420	158,252	301,939
United States	35+	11,378	5,630	4,737
South America				
Argentina	30–40	25,954	24,095	24,000[b]
Bolivia[e]	35+	–	–	93
Brazil	38–50	1,852,000	2,166,000	2,126,000
Chile	41–47	26,283	26,124	29,457
Guyana	36–42	144,138	–	–
Peru	30+	7,885	13,228	2,119
Europe				
Bulgaria	30+	45,000	43,000	44,000[b]
Greece	50	7,434	7,125	7,190
Hungary	30–	172,000	172,000	186,028
Italy	30–	56,020	58,385	55,216
Portugal	38–44	10,654	7,637	6,083
Spain	30+	14,248	25,302	11,504
USSR[f]	g	7,236,000	7,221,000	7,700,000[b]
Yugoslavia	30+	15,582	13,593	16,298
Africa				
Angola	30+	10,086	32,044	25,000
Botswana	30+	11,021	24,769	45,020
Congo (Kinshasa)	42+	354,735	343,291	382,446
Gabon	50–53	1,382,958	1,502,000	1,601,700
Ghana[h]	48+	455,617	366,800	446,837
Ivory Coast	32–47	128,685	140,036	25,419
Morocco	35–53	176,602	143,935	123,873
Republic of South Africa	30+	2,173,438	2,429,600	2,953,609
Sudan	36–44	5,500	940	1,279
United Arab Republic	30–	4,361	4,400	g
Zambia	35+	27,999	28,284	33,000[b]

electrolytically or chemically, the electrolytic process being very similar to that for the production of electrolytic metal. Manganese dioxide is used here for the readily available oxygen it contains, rather than for its manganese content.

The same, and other, manganese dioxide ores are used as oxidants in the production of hydroquinone, in the leaching of uranium ores, in the electrolytic production of zinc, and in various chemical processes. They are used in the production of potassium permanganate, manganese sulfate, manganous oxide, manganous chloride, and other manganese

Manganese in the Ecosystem

TABLE 2-3 (*Continued*)

Country	% Manganese[b]	Ore Production, tons		
		1968	1969	1970[c]
Asia				
China, mainland[b]	30+	990,000	1,100,000	1,100,000
India[i]	[g]	1,766,000	1,637,000	1,820,000
Indonesia	35-49	2,400	7,100	2,200
Iran[j]	35+	28,000	39,000	40,000[b]
Japan	30-43	344,247	331,587	298,701
Korea, Republic of	35+	4,653	3,199	3,749
Malaysia	30-40	49,737	11,392	—
Philippines	30-52	72,800	22,048	5,645
Thailand	40+	45,270	32,872	26,307
Turkey	30-50	27,842	15,090	10,465
Oceania				
Australia	46	819,692	1,016,286	886,080
Fiji	30-50	9,750	22,917	27,054
New Hebrides	31-43	46,824	—	16,926
TOTALS		18,628,213	19,195,961	20,389,974

[a] Derived from DeHuff.[130] In addition to the countries listed, Cuba and Southwest Africa produce manganese ore, but information is inadequate to permit reliable estimates of output. Colombia has produced ore of unspecified grade, in the following amounts (in tons): 1968, 551; 1969, 606; 1970, 511. Low-grade ore, not included in table, has been produced in the following countries (all amounts in tons): in Argentina (about 22% Mn): 1968, 11,210; 1969, 16,151; 1970, 17,000 (estimated); in Czechoslovakia (about 17% Mn): 1968, 95,000; 1969, 93,000; 1970, 94,000 (estimated); in Rumania (about 22% Mn): about 140,000 each year; in Republic of South Africa (15-30% Mn): 1968, 502,080; 1969, 484,041; 1970, 412,264; in Sweden (about 12% Mn): 1968, 12,921; 1969, 9,700.
[b] Estimated.
[c] Preliminary.
[d] Estimated on basis of reported contained manganese.
[e] Estimated on basis of exports.
[f] Grade unreported.
[g] Not available.
[h] Dry weight.
[i] Not reported by grade; of total exports of 1,310,769 tons in 1968 and 1,331,349 tons in 1969, 61% and 57%, respectively, graded less than 35% Mn.
[j] Based on Iranian calendar year, beginning March 21; all figures refer to mine-run ore.

chemicals. Potassium permanganate is a strong oxidant used for the purification of public water supplies, as well as other uses. Manganese sulfate and manganous oxide are used to add manganese to manganese-deficient soils or to soils that do not readily give up their contained manganese—either as a trace element in fertilizers or by direct application. The Florida citrus belt is a principal beneficiary. Manganous oxide is the preferred form for the introduction of manganese as a supplement in animal and poultry feeds. Manganous chloride is the principal constituent of a flux added to magnesium to give it hardness, stiffness, and corrosion resistance. It is also the starting point in the production of the fuel additive, methylcyclopentadienyl manganese tricarbonyl.

Manganese ores, alloys, metal, and chemicals are used in welding-rod coatings and fluxes. Manganese ores and chemicals made from them are used to produce various color effects in face brick and, to a much smaller extent, to color or decolor glass and ceramic products. They are of various degrees of importance in producing dyes, paint and varnish dryers, fungicides, and pharmaceuticals. Manganese dioxide is used as a constituent of the sealant in the base of incandescent light bulbs[147] and as a constituent of frits for bonding porcelains and glass to metal. Both manganese oxides and powdered electrolytic manganese metal are used to produce the manganese-zinc-ferrites used in magnets for various electronic applications.

The consumption of manganese in the United States for its various uses in 1970, as reported to the U.S. Bureau of Mines, is shown in Tables 2-4 and 2-5.

Emission to the Atmosphere

Qualifying factors that require attention in considering manganese emission to the atmosphere include geographic location (including topography), chemical composition, size and shape of particles, distance of sample from emission source, height above ground, quantity and nature of other unrelated pollutants in the atmosphere, wind, other atmospheric conditions[165,326] at the time of sampling, duration of air sampling, and method of sampling and analysis. Pollution from industrial plants will

TABLE 2-4 Consumption of Manganese Ore in the United States in 1970[a]

Use	Gross Weight of Ore,[b] tons
Manganese alloys and metal	2,099,426
Pig iron and steel	107,733
Dry cells, chemicals, and miscellaneous	156,778
TOTAL	2,363,937

[a] Derived from DeHuff.[130]
[b] By definition, manganese ore contains at least 35% manganese (natural).

TABLE 2-5 Consumption of Manganese Ferroalloys and Metal in the United States in 1970 by End Use[a]

	Gross Weight, tons				
	Ferromanganese				
Use	High-Carbon	Medium- and Low-Carbon	Silico-manganese	Spiegeleisen	Manganese Metal[b]
Steels					
Carbon	713,681	91,550	87,845	11,169	4,837
Stainless and heat-resisting	1,830	4,931	8,371	c	6,689
Alloy (excluding stainless and tool)	108,043	27,539	26,852	1,679	1,960
Tool	839	122	c	–	c
Cast irons	8,471	2,343	7,154	7,309	11
Superalloys	437	44	c	–	363
Alloys (excluding alloy steels and superalloys)	5,387	1,172	1,838	–	9,410
Miscellaneous and unspecified	32,423	1,799	6,440	111	1,212
TOTALS	871,111	129,500	138,500	20,268	24,482

[a] Derived from DeHuff.[130]
[b] Nearly all electrolytic.
[c] Withheld to avoid disclosing individual company confidential data, but included in "Miscellaneous and unspecified."

tend to be localized, whereas other pollution from the combustion of fuels will have much wider distribution. Conversely, the effects from industrial plants can be much more serious in their restricted areas, both inside and outside a plant, with the severity usually varying inversely with distance from the source.

In view of these observations, data that are available for any study of manganese emission to the atmosphere and of manganese in the atmosphere must be selected and used carefully.

In this discussion, attention will be directed primarily to emissions to the general atmosphere, rather than to localized in-plant emissions, which must, in effect, be treated as individual case studies because of the many variables involved. In following through on the general approach, one finds that the data are limited or even lacking entirely, seldom representative, and often not in agreement. Seldom are they on readily comparable bases. Much of the information that is available is of questionable value because of its age, and the literature at times fails to define clearly the degree of control of the emissions being described. Most metallurgic data appear to be for emissions before cleaning;

unless otherwise stated, the following emission data can be considered to be on that basis.

Fume has been defined as the product of vaporization and later condensation, producing particles of 0.1–1.0 μm, whereas dust is generally considered to consist of mechanical dispersoids, such as those created by grinding.

By the grinding process it is difficult to obtain particles smaller than 5 microns in size. This exceedingly large difference in size between fume and dust particles is most readily contrasted by the fact that a 10 micron dust particle has 8000 times the volume and weight of a ½ micron fume particle, both having the same composition.[242]

In metallurgic emissions, however, the two usually are intimately mixed and the terms are often used indiscriminately. Person[400] places the cut-off point between dust and fume at approximately 2 μm.

In the case of emissions from metallurgic furnaces, volume and manganese content depend to a considerable extent on the physical nature and the chemical composition of the furnace feed.

MANGANESE ALLOY PRODUCTION

A principal source of man-made manganese pollution of the atmosphere on a unit basis in the past has been the ferromanganese blast furnace, but quantitative data on present emissions are limited or lacking.

The dust found in ferromanganese blast furnace gas is of two types. One type, comprising about 20 percent of the dust present, consists of particles above 20 microns in size which, from their analysis, appear to originate in the disintegration of the coke and ferromanganese grade ore in the furnace burden. This fraction of the dust is removed from the gas stream by a conventional dust catcher. The other type, which accounts for 80 percent of the solids present in the gas as it leaves the furnace, is a typical fume, with particle sizes varying from 0.10 to 1.0 micron. This material is apparently formed by a process of vaporization and condensation, which explains its extremely small size. It is the presence of this material which renders the problem of cleaning ferromanganese blast furnace gas so vastly different from the cleaning of basic blast furnace gas. [A number of samples of the fume showed a manganese content of 15–25%, an apparent density of 192 kg/m^3, and an average particle size of 0.3 μm.] "On the basis of many tests, it was found that the fume loadings averaged about 7.5 grains per standard cubic foot. . . . The fume was also found to be pyrophoric. If heated to approximately 350 F and exposed to air, it begins to oxidize, with the liberation of considerable heat, slowly turning from a gray to brown color as the oxidation proceeds. During this color change, the manganese oxides originally present are further oxidized and the carbonates present are decomposed. This burning operation also increases the density of the fume

Manganese in the Ecosystem

and renders it less floury. The data taken from a study of the gas moisture content and temperature variations in the blast furnace gas to be cleaned showed that the variation in the moisture content ranged from a low of 2 percent to a high of 30 percent, and that the changes in moisture content were usually very rapid. . . .

A pilot plant was designed to clean 3000 standard cubic feet per minute of dry ferromanganese blast furnace gas in order to obtain sufficient clean fuel to fire one of the boilers in the plant. This represents about one-fifteenth of the gas produced by one blast furnace. Such an arrangement permitted a study to be made of the advantages to be gained at the boiler, made possible a visual guide as to the degree of gas cleaning attained, and provided approximately 190 pounds per hour of fume for further processing. The pilot plant operation showed a visible effect of gas cleaning on the boiler stack effluent which was impressive. The stack of the boiler in which the cleaned gas was burned did not emit a discernible fume, while the others did. Although the loading of the cleaned gas entering the one boiler was about 0.12 grains per standard cubic foot, the stack effluent loadings were only about 0.05 grains per standard cubic foot because of the dilution of the gas with the air used for combustion. The cleaned gas burned with a short bright blue flame.[60]

Operation of this pilot plant was followed by installation and successful operation of full-scale equipment to clean 135,000 cubic feet per minute (cfm) of gas produced daily by two ferromanganese blast furnaces at the Duquesne Works of United States Steel Corp. Approximately 105 tons of dust per day were collected by the electrostatic precipitators used.[203] This installation apparently was responsible for the statement made in 1958 by the Director of the Allegheny County Bureau of Smoke Control, Pittsburgh, Pa., that electrostatic precipitators collected more than 100 tons of particulate matter per day from two furnaces having a daily production of approximately 670 tons of alloy.[554] This is roughly 150 tons of dust per 1,000 tons of alloy.

In 1970, there were only three blast furnaces making ferromanganese in the United States; all were in western Pennsylvania—two in the Pittsburgh district (Clairton and McKeesport) and one at Johnstown. In the previous year or two, blast-furnace ferromanganese was made also at Birmingham, Ala., and at Sheridan, near Reading in eastern Pennsylvania. It should be borne in mind, moreover, that the pig-iron blast furnace can be used to make either pig iron or ferromanganese; this interchangeability option is exercised at times, although not often.

In 1970, there were 11 domestic plants regularly producing ferromanganese, silicomanganese, or both in electric furnaces. They were in or near Philo, Ashtabula, Marietta, and Beverly, Ohio; Calvert City, Ky.; Niagara Falls, N.Y.; Houston, Tex.; Memphis and Rockwood, Tenn.; Portland, Oreg.; and Alloy, W. Va. Although individual capacities of the electric furnaces were considerably less than those of the largest blast furnaces, their total ferromanganese production was greater in 1970. In previous years, the blast-furnace total was larger, usually in a ratio of approximately 2:1. Total ferromanganese production in the United States in 1970, as reported to the Bureau of Mines, was 835,000 tons.[130] Individual practices vary, but, as a general rule, it can be figured that it takes roughly 2 tons of manganese ore to make 1 ton of ferromanganese.

Manganese content and particle size of fume from various electric ferroalloy furnaces are shown in Table 2-6. The stoichiometry involved in producing a particular alloy and the type of furnace used are important variables that affect the volume of gas generated. Furnace gas generation, without combustion (as in the case of a covered furnace), is given *approximately* as 160-170 standard cubic feet per minute (scfm) per megawatt for standard ferromanganese, and 120-130 scfm for silicomanganese, but momentary surges or peak flows can increase the volume as much as 40%.[400]

"When the furnace gas burns with air, as with an open furnace, a significant volume increase occurs, as high as a factor of 50 depending on the amount of induced air, in the volume to be treated for dust collection." An average measured efficiency of 98.6% was obtained by wet dust collectors installed on a low-hood silicomanganese open furnace. The inlet loading to these wet scrubbers was 1.31 grains/scf, and the outlet loading was 0.017 grains/scf. These and other details are set forth in Table 2-7. "The color visibility threshold for manganese or silicomanganese fume has been judged to be in the range of 0.02 to 0.03 grains per standard cubic foot." Bulk density of collected electric-furnace fume in the dry state is very light, at about 2-14 ka/m^3.[400]

Dusts recovered by tubular centrifuges, of 20-50% efficiency, from French electric ferromanganese and silicomanganese furnaces, reported in 1960, had manganese contents of 20-35%.[364]

A classic example of manganese pollution of the atmosphere by ferroalloy furnaces was that of some 40 years ago at Sauda, Norway, where large quantities of thick brown smoke were emitted to the atmosphere by open-top electric-arc furnaces in the course of producing ferromanganese and silicomanganese. Analyses made in 1926 of the dry substance of this smoke obtained at the point of emission from the furnaces showed it to contain approximately 30% manganese oxides.[151] Because of the furnace location at the bottom of a narrow fjord, in a humid climate subject to many windless and rainy days, the smoke often remained as a blanket over the district. Analyses of the smoky air collected at several points in the community in 1930 showed a particle

TABLE 2-6 Typical Furnace Fume Characterizations[a]

Furnace Product	50% FeSi	SMZ[b]	SiMn[c]	SiMn[c]	FeMn	H.C. FeCr	Chrome Ore–Lime Melt	Manganese Ore–[c] Lime Melt
Furnace type	Open	Open	Covered	Covered	Open	Covered	Open	Open
Fume shape	Spherical, sometimes in chains	Spherical, sometimes in chains	Spherical	Spherical	Spherical	Spherical	Spherical and irregular	Spherical and irregular
Fume size, μm								
Maximum	0.75	0.8	0.75	0.75	0.75	1.0	0.50	2.0
Most particles	0.05–0.3	0.05–0.3	0.2–0.4	0.2–0.4	0.05–0.4	0.1–0.4	0.05–0.2	0.2–0.5
X-ray diffraction trace constituents[d]	FeSi FeSi$_2$	Fe$_3$O$_4$ Fe$_2$O$_3$ Quartz SiC	Mn$_3$O$_4$ MnO Quartz	Quartz SiMn Spinel	Mn$_3$O$_4$ MnO Quartz	Spinel Quartz	Spinel	CaO
Chemical analysis, %								
SiO$_2$	63–88	61.12	15.68	24.60	25.48	20.96	10.86	3.28
FeO	—	14.08	6.75	4.60	5.96	10.92	7.48	1.22
MgO	—	1.08	1.12	3.78	1.03	15.41	7.43	0.96
CaO	—	1.01	—	1.58	2.24	—	15.06	34.24
MnO	—	6.12	31.35	31.92	33.60	2.84	—	12.34
Al$_2$O$_3$	—	2.10	5.55	4.48	8.38	7.12	4.88	1.36
LOI	—	—	23.25	12.04	—	—	13.86	11.92
TCr as Cr$_2$O$_3$	—	—	—	—	—	29.27	14.69	—
SiC	—	1.82	—	—	—	—	—	—
ZrO$_2$	—	1.26	—	—	—	—	—	—
PbO	—	—	0.47	—	—	—	—	—
Na$_2$O	—	—	—	2.12	—	—	1.70	0.98
BaO	—	—	—	—	—	—	—	2.05
K$_2$O	—	—	—	—	—	—	—	1.13
	—	—	—	—	—	—	—	13.08

[a] Derived from Person.[400]
[b] Si, 60–65%; Mn, 5–7%; Zr, 5–7%.
[c] Manganese fume analyses in particular are subject to wide variations, depending on the ores used.
[d] All fumes were primarily amorphous.

TABLE 2-7 Data on Wet Scrubbers for a Low-Hood Silicomanganese Open Electric Furnace[a]

Original completion date	1968
Furnace rating for collector design	30 MW
Measured dust-collection efficiency, average	98.6%
Inlet loading	1.31 grains/scf
Outlet loading	0.017 grain/scf
Design volume at furnace hood	255,000 acfm
Design temperature	620 F
Actual duct or offtake temperature	490–550 F
Design volume handled by fans or blowers	196,000 acfm
Operating pressure drop across collector	57 in. H_2O
Installed horsepower	
fans	2,800 hp
auxiliaries	50 hp
Dust collected	14–17 tons/day
Uncollected emission	0.20 ton/day
Water circulation	1,800 gpm
Water usage	350 gpm

[a] Derived from Person.[400]

size below 5 μm. Near the plant, the smoky air had a manganese oxide content of 2.6% and a high silica content of 54%.[150] "These analyses were not made on days with a particularly high ratio of smoke in the air. We can therefore assume that the amounts will be more on smoky days and less on windy days."[150] Riddervold, a physician resident of the area, summarized the data as follows:

The analyses we have to go by were carried out by Dr. Böckmann in 1930, and it is striking how little manganese there really is in the air—1 m^3 of air never contains more than 0.1 mg manganese oxide. It turns out that the dust suspended in the smoke contains 36% to 38% manganese oxide. 90 m farther away, at the so-called office building, the same dust contains only 2.5% to 6% manganese oxide, and when we come up to the first building, approximately another 100 m from the office building, the manganese oxide content in the dust is found to be 0.17% to 0.18%.[418]

However, Böckmann[62] commented that the method of analysis that he used was later found to be unreliable, suggesting that the values obtained were too low. He expressed, as follows, his analyses of smoke taken in the spring of 1930, apparently the same analyses referred to above, in terms of milligrams of manganese tetroxide (Mn_3O_4) per cubic meter:

In the foreman's residence (relatively close to the mill): 0.015, 0.092, and 0.055 ($Mn_3O_4:SiO_2$ ratios: 1:158, 1:100, and 1:80)

In the school (a little farther from the mill): 0.023, 0.011, 0.029, and 0.019 ($Mn_3O_4 : SiO_2$ ratios: 1:270, 1:127, 1:286, and 1:237)

In Sauda tourist hotel (approximately 3 km from the mill): 0.030 and 0.064 ($Mn_3O_4 : SiO_2$ ratios: 1:295 and 1:99)

The manganese tetroxide:silica ratio for smoke collected in the office building of the mill was 1:21. From Böckmann's table,[62] it appears that a total of nine samples were taken, each on a different day. With the question of validity of the method of analysis, the data are of little value, except to say that at least the stated quantities of manganese oxide and silica were in the smoke, or smoky air, at the time and place sampled.

STEEL AND IRON PRODUCTION

Steel-furnace dusts show considerable variation in chemical composition, depending on a number of factors, including composition of the charge, operating procedures, and the point in the operating cycle at which the sample is taken. In a 1970 study, the Bureau of Mines obtained the following average manganese analyses of 340 dust samples from 48 operators: electric furnaces, 4%; open hearths, 0.4%; and basic oxygen furnaces, 1%. A survey of 95 steel mills at that time showed the following average dust emissions, before cleaning, in pounds per ton of steel produced: electric furnaces, 20; open hearths, 25; and basic oxygen furnaces, 40.[41]

Data compiled in 1957 for electric-arc furnaces making steel, at both steel mills and steel foundries, showed dust and fume emissions ranging from about 2 to 13 kg/ton of steel.[76] Size analysis of a typical sample had been reported earlier as follows:[156]

size, μm	distribution, %
0-5	67.9
5-10	6.8
10-20	9.8
20-44	9.0
44 and over	6.5

One foundry operator, running two 6-ton furnaces, collected 3 tons of dust every 24 h. With 40,000 cfm passing through the collecting system, the average dust loading was 0.7 grains/ft^3.[156] Dust generation in large electric furnaces—e.g., steel-mill furnaces of 75- and 50-ton capacity—might run as high as about 506 kg/h and about 14 kg/ton of total charge. With a typical charge consisting of 5% fluxes, carbon, and ore, and the remainder consisting of scrap of various kinds, the typical dust emission contained 4% manganese oxide.[117]

The open-hearth furnace generates a "large volume of waste gases and dusts."[326] Chemical analysis of a composite sample of dust collected some years ago during all the periods of an operating cycle for an open-hearth steel heat showed a manganous oxide content of 0.63%; 46% of the particles were smaller than 5 μm, and dust loading was 0.1–1.6 grains/scf of dry gas.[61]

The fast-growing basic oxygen process, rapidly replacing the open hearth, has some similarity to the historic Bessemer process, in that both are blowing operations that generate large volumes of dust and fume. However, all basic oxygen furnaces are being built with efficient equipment to control emissions. Cleaned gases emitted to the atmosphere contain solids at less than 0.05 grain/ft^3, and the discharged cleaning water has less than 5 grains of settleable solids per gallon[338 (p.490)] A typical analysis of the dust and fume, after ignition of the gases on meeting air at the vessel's mouth, shows 4.4% manganese tetroxide and more than 90% iron oxides, which have the following particle size distribution:

size, μm	wt. %
<0.5	20
0.5–1.0	65
1.0–15	15

Tobacco smoke, by comparison, has a particle size of approximately 0.5 μm. The total weight of the particles emitted by a basic oxygen furnace varies from 1% to 2% of the weight of the metallic charge. The dust loading, determined after combustion and dilution of the gases, ranges from zero to 20 grains/scf during the course of a heat. For purposes of designing the cleaning system, an average value of 7–9 grains/scf is normally used. This means that a minimal gas-cleaning efficiency of 99.3–99.4% is required in order to achieve the desired 0.05 grain/scf.[189] Guaranteed efficiencies of 99.5% and 99.7% have been reported for dry electrostatic cleaning systems, and a discharge of 0.004 grain/ft^3 has been reported for a wet disintegrator system in which the uncleaned gases had a load of 2.8–19.3 grains/ft^3 at a temperature of approximately 16 C.[511]

The Bessemer process is no longer used in the United States, and the Thomas (or basic Bessemer) process was never adopted in this country. The Thomas process, however, was and still is an important steel-making process in Europe,[338] being particularly adaptable to iron

Manganese in the Ecosystem

ores of high phosphorus content. The dusts arising from the handling and grinding of Thomas slags (used for fertilizer) contain manganese, variously reported (or translated) as 6–8% manganese[150,269,270] or 6–8% manganese oxides.[151,268,269] These dusts were recognized many years ago as an occupational hazard because of the predisposition to pneumonia by their inhalation.[151]

Total U.S. raw steel production in 1970 was 131.5 million tons—20.2 million tons in electric furnaces, 63.3 million tons in basic oxygen furnaces, and the remaining 48 million tons in open hearths.[12] The proportion of manganese in the dusts and fumes emitted in the production of steel will be a function of the furnace feed, rather than the process used, except as the latter has a bearing on the former. Volume, however, will be directly related to the process used, as well as to the physical character of the feed and the efficiency of the pollution-control equipment.

Before any cleaning, the pig-iron blast furnace may emit from about 38 kg[338 (p.457)] to approximately 90 kg[225] of dust for each ton of iron produced. The smaller quantity reflects an iron-bearing burden consisting of approximately 1,110 kg of fluxed sinter, 276 kg of iron ore, and 89 kg of scrap. In the case of the larger and older figure, apparently an empirical one, it can be assumed that the quantity of sinter, nodules, or pellets in the charge would be appreciably less than the quantity of crude iron ore or concentrate. This assumption is supported by the following earlier data from another source covering 10 days of operation of a 300-ton Alabama furnace in or before 1927: For each long ton of iron produced, 126 kg of flue dust were obtained with an iron-bearing burden of 451 kg of nodules, 1,450 kg of iron ore, and 149 kg of scrap. The flue dust was analyzed at 0.49% manganous oxide.[282] These differences in the quantities of flue dust emitted can probably be attributed largely to the much higher content of fines in iron ores than in sinter.

The use of primary gas cleaning is almost universal practice in the operation of blast furnaces and in the most effective apparatus, the gas leaves the primary washer with a dust content of 0.1 to 0.3 gr./ft.^{3} Blast furnace gas fine-cleaned in modern secondary gas cleaning apparatus has a dust content of no more than 0.01 gr./ft.^{3} and as a result the dust emitted to the open atmosphere with the burned waste gases amounts to less than 10 lb/hr/furnace. For all practical purposes this is insignificant. Nearly all of the blast furnace gas in the U.S. is cleaned in primary gas washers and a major portion also is fine-cleaned in secondary washers or precipitators.[225]

That was written in 1957. Particle size of the dust was reported somewhat earlier to range from about 0.64 cm down to 0.1 μm.[242] In spite of the cleaning equipment, there will be occasional escapes of uncleaned flue dust due to irregularities in operation caused by hanging of the charge, or "slips," in the furnace. The economics of pig-iron production demand that these be avoided insofar as possible, and the increased use of pellets in the iron burden has been a big factor in decreasing the frequency of "slips."

Pig-iron production in the United States in 1970 totaled approximately 91 million tons.[12]

The fumes and dust from iron and steel foundries can be assumed to have manganese contents comparable with those from the various steel furnaces, or possibly somewhat higher in some cases, with varying degrees of efficiency of volume control. Even with no control, individual plant emissions and total emissions from foundries would be much less than those from steel mills, because foundries are much smaller operations. However, the foundries are widespread, and uncontrolled emissions could create numerous local pollution problems.

The foregoing discussion has considered primarily fume and its associated dusts. All the above operations, however, will have dusts arising from the handling, crushing, sizing, mixing, and sometimes storage of manganese ores and other manganiferous materials (including recovered dusts). Conventional techniques are available for handling the various dusts at these plants, and with proper application they are not ordinarily considered to be a pollution problem.[400] Hence, they may be considered to represent essentially a local problem that depends on the degree of control exercised. The problems that attend the disposition or storage of recovered fine dusts and their potential for creating pollution problems are not to be ignored, however.

OTHER SOURCES OF EMISSION

The comments concerning dusts will apply to the other industrial operations in which manganese-bearing dusts arise, chiefly from the handling of materials. These include mining operations; concentrating and blending plants at mines or ports of entry; loading and unloading operations related to ships, railcars, and trucks; dry-cell battery manufacture; synthetic manganese dioxide production; manganese metal production; production of chemicals; production and use of welding rods and welding-rod coatings; and mixing and application of fertilizers, fungicides, and animal feeds.

There is currently no domestic mining of manganese ore, defined as containing 35% or more manganese. Mining of lower-grade manganiferous ores is confined to the Cuyuna Range of Minnesota and to southwestern New Mexico. These ores, as shipped presently, contain about 10-20% manganese and 30% or more iron.

There are crushing, concentrating, or blending plants handling imported ores at or near Newport News and Lynchburg, Va.; Philadelphia, Pa.; York, Ala.; Brownsville and El Paso, Tex.; Wilmington, Del.; and Camden, N.J.

Dry-cell battery plant locations include (or until recently have included) Red Oak, Iowa; Greenville, Asheboro, and Charlotte, N.C.; Fremont, Cleveland, and Lancaster, Ohio; Bennington, Vt.; Stamford, Conn.; Clifton, N.J.; Freeport, Ill.; Wausau, Madison, Fennimore, Wonewac, and Appleton, Wis.; St. Paul, Minn.; Williamsport, Pa.; Waco, Tex.; Marysville, Mo.; and Woodruff, S.C.

Plants for the production of synthetic manganese dioxide or various manganese chemicals are at Marietta, Ohio; Baltimore, Md.; Henderson, Nev.; La Salle, Ill.; Kingsport, Tenn.; Bowmanstown, Pa.; and Jersey City, N.J. Until recently, there was one at Covington, Tenn.

Electrolytic manganese metal plants are at New Johnsonville, Tenn.; Aberdeen, Miss.; and Marietta, Ohio. Until recently, there was one at Knoxville, Tenn.

No attempt will be made to list the various miscellaneous users of manganese ores and other manganiferous materials, some of which might be sources (probably minor) of manganese emissions to the atmosphere, as may also be the incineration of manganese-bearing wastes.

Small quantities of manganese apparently are present in nearly all U.S. coals, as shown by ash analyses of 781 samples ranging from 0.001% to 0.3% manganese.[3] Fly-ash collection efficiencies vary for coal-fired power plants, but efficiencies of more than 95% are not uncommon. Particulate emissions for two plants tested in a joint study by the Public Health Service and the Bureau of Mines in about 1964 had dust loadings ranging from 0.13 to 0.74 grain/scf after passing the cleaning equipment.[194] Semiquantitative analysis by emission spectroscopy showed the manganese emission after cleaning to range from 0.28×10^{-4} to 1.2×10^{-4} grain/scf, with an estimated accuracy within 50%. The average metal-collection efficiency for one plant, having an electrostatic precipitator following a cyclone separator, was 93%; that for the other plant, having a cyclone separator but no electrostatic precipitator, was 70%. In the first plant, 85% of the fly ash after the collector was larger than 3 μm; in the second plant, the proportion was 73%.[194]

Fly ash from oil-burning boilers under investigation was reported to contain 0.04% manganese as manganese dioxide. The normal range of dust loading is stated to be 0.025-0.060 grain/scf; the extreme range,

0.005–0.205 grain/scf; and the most commonly reported values, 0.030–0.035 grain/scf. Particle size apparently ranges from less than 0.01 to 20 μm and up, with size distribution depending on the degree of atomization of the oil, mixing efficiency, flame temperature, firebox design, flue-gas path through the boiler to the stack, and number of collisions between fly-ash particles.[460]

MANGANESE IN THE ATMOSPHERE

Sulfur dioxide emitted to the atmosphere in the presence of oxygen, sunlight,[191] and water vapor is converted to sulfur trioxide and then to sulfuric acid.[169] The presence of ammonia or some metal compounds, including those of manganese, is capable of promoting this reaction.[116, 159,333,342] The unanswered questions seem to be related to the degree of concentration of manganese required to do this and the significance of its effect. If manganese concentrations that might be expected in the atmosphere can be responsible for the generation of a significant quantity of sulfuric acid that would not have been formed otherwise, then manganese can cause a health hazard on this score, apart from any hazard it might constitute in itself. Opinions differ as to whether the manganese concentrations could be large enough to make this a matter of concern. The available evidence seems to support the contention that a higher manganese concentration would be necessary.

Although manganese oxides act as catalysts, interest in the use of manganese oxides to remove sulfur dioxide from flue gases emanating from power plants, etc., appears to have been primarily from the standpoint of their use as absorbents, rather than catalysts. They have been applied as absorbents, with the attendant formation of manganese sulfate, in Japan.[273] Although it is possible that manganese oxides have also been used primarily as catalysts for removal of sulfur oxides, vanadium catalysts are favored for this purpose.

Manganese dioxide reacts with nitrogen dioxide in the laboratory to form manganous nitrate,[442] and such a reaction might occur in the atmosphere after emission of manganese. Apparently, both manganous nitrate and nitrogen dioxide are toxic,[437] and it is conceivable that the net result of such conversion, if it does occur, in terms of pollution of the atmosphere would be little change or even a change for the better.

In 1953, the Public Health Service, in cooperation with some state and local agencies, initiated an air sampling program in 17 U.S. cities. By 1956, coverage had been expanded to 66 communities, but a truly national network was not established until January 1957, when some 185 urban and 48 nonurban stations were engaged in the program. Because manganese represents only one of many determinations to be made from the samples collected in the program, and others have been and are of greater interest to pollution-control efforts, there has been and continues to be a great variation in the lag in analyzing for manganese and reporting the results.

Analysis is by semiquantitative emission spectrography, with the minimal detectable manganese concentration being 0.01 $\mu g/m^3$ for urban samples and 0.006 $\mu g/m^3$ for nonurban, it having been found possible to obtain greater sensitivity for the nonurban analyses.[501,510]

Some samples are analyzed individually; others are analyzed as quarterly composites. Beginning January 1, 1957, National Air Surveillance Networks (NASN) stations have operated throughout 26 randomly selected 24-h days in a year.[501,510] Other contributing stations may operate a greater or smaller number of days in a year.[505] Some cities have reports from several different sampling sites. The NASN urban stations are in central business–commercial districts. Although they are not representative of the cities' average atmosphere, they are considered to permit valid comparisons with NASN stations in other cities.[510] It is obvious that reports of exceptionally high maximal concentrations are of limited value without firsthand knowledge of station locations and the particular atmospheric conditions at the times of sampling. Also, comparisons of data between different years involved some problems because of various changes in sampling, analytic methodology, and method of reporting over the years, but especially because the stations represented in the various reports are not necessarily the same ones each year.

The Public Health Service report for 1953–1957[502] showed that 12 nonurban samples, all collected in 1955–1956 at Pt. Woronzof, Alaska, averaged 0.01 μg of manganese per cubic meter, with a maximum of 0.02 $\mu g/m^3$; 140 suburban samples collected at nine different locations in the conterminous United States in 1954–1956 averaged 0.06 $\mu g/m^3$, with a maximum of 0.50 $\mu g/m^3$ (Kanawha County, W. Va.); and 1,962 urban samples collected in 1953–1957, but mostly in 1954–1956, averaged 0.11 $\mu g/m^3$, with a maximum of 9.29 $\mu g/m^3$ (at Cincinnati, Ohio, in 1955). It is interesting to note that the third highest maximal analysis was 3.74 $\mu g/m^3$, for Anchorage, Alaska (1954–1955). This appears to have been the result of high concentrations of particulate matter that occurred in the atmosphere after a volcanic explosion. It followed closely on Philadelphia's 3.86 $\mu g/m^3$ (1954). Chattanooga, Tenn., produced the fourth highest, 3.00 $\mu g/m^3$ (1955–1956). These early data for 1953–1956 were obtained by sampling and analytic procedures entirely different from those used after 1956, but they are considered to represent the correct order of magnitude. However, the aforementioned averages are of value only as orders of magnitude.

The report for 1957–1961[504] gives 0.10 $\mu g/m^3$ as the arithmetic mean for 819 urban samples collected about the nation in 1957–1960. The maximum was 2.60 $\mu g/m^3$, for Cleveland, Ohio, in 1958, followed by 2.20 $\mu g/m^3$, for Pittsburgh, Pa., in 1960. However, in another tabulation, for a different group of stations, reporting high samples on a quarterly basis, Johnstown, Pa., had highs of 7.80 and 5.70 $\mu g/m^3$ for two quarters of 1959; Charleston, W. Va., had a high of 7.10 $\mu g/m^3$ in one quarter of 1958; Gary, Ind., was high for each quarter of 1960, with 2.20, 3.10, 2.00, and 3.00 $\mu g/m^3$; Canton, Ohio, in 1959 had a high of 2.20 $\mu g/m^3$ for one quarter; and Oakland, Calif., in 1958 had one quarter high at 2.70 $\mu g/m^3$.

The 1962 report[506] does not attempt any national averages but gives the following urban-station maxima not previously reported: Charleston, W. Va., 9.98 $\mu g/m^3$ in 1961 and 4.70 $\mu g/m^3$ in 1959; Philadelphia, Pa., 9.98 $\mu g/m^3$ in 1961; Canton, Cincinnati, and Columbus, Ohio, 2.90, 2.40, and 1.50 $\mu g/m^3$, respectively, in 1961; and Youngstown, Ohio, 2.10 $\mu g/m^3$ in 1960.

Analyses reported as 9.98 in the above and following series were in fact greater than 10, but are considered to be of that order of magnitude.

The report for 1963[505] has the following urban-station maxima: Cincinnati, Ohio, 2.30 $\mu g/m^3$ in 1962 and 1.80 $\mu g/m^3$ in 1960; Philadelphia, Pa., 1.50 $\mu g/m^3$ in 1960 and 1.30 $\mu g/m^3$ in 1962; and Columbus, Ohio, 1.30 $\mu g/m^3$ in 1962.

The 1964–1965 report[507] gives 0.10 $\mu g/m^3$ as the arithmetic average of 103 urban-station-year entries, including those analyzed on an individual-sample basis and those on a quarterly-composite basis, representing a total of approximately 70 cities. The samples were collected in 1960–1964. Maxima were reported only for cities analyzed on an individual-sample basis. Charleston, W. Va., had the highest, with 9.98 $\mu g/m^3$ in 1964; Philadelphia, Pa., 3.70 $\mu g/m^3$ in 1963 and 1.90 $\mu g/m^3$ in 1964; Cincinnati, Ohio, 2.00 $\mu g/m^3$ in 1964 and 1.30 $\mu g/m^3$ in 1963; Cleveland, Ohio, 1.70 $\mu g/m^3$ in 1962, 0.96 $\mu g/m^3$ in 1963, and 1.00 $\mu g/m^3$ in 1964; and Pittsburgh, Pa., 1.20 $\mu g/m^3$ in 1962, 1.40 $\mu g/m^3$ in 1963, and 0.82 $\mu g/m^3$ in 1964.

Urban quarterly averages reported for 1965 collections in the 1966 report[501] ranged from below the minimum detectable for two quarters at Brockton, Mass., to 0.31 $\mu g/m^3$ at Birmingham, Ala., and 0.25 $\mu g/m^3$ at Bethlehem, Pa.; yearly averages ranged from Brockton's low of 0.01 $\mu g/m^3$ to highs of 0.15 $\mu g/m^3$ for Birmingham, Ala., 0.13 $\mu g/m^3$ for

Akron, Ohio, and 0.12 µg/m³ for Bethlehem. The arithmetic average of urban-sample averages tabulated in this 1966 report, including the above 1965 samples and samples for earlier years (mostly 1963 and 1964), is 0.09 µg/m³, representing 55 cities. Maxima were reported only for the stations analyzed on an individual-sample basis. Of these, Johnstown, Pa., was high, with 6.90 µg/m³ for 1963. Portland, Oreg., at 1.30 µg/m³ for 1964, was the only other high maximum.

The next, most recent report[510] has only one urban-manganese table (Table 2-8), a tabulation for stations analyzed on an individual-sample basis. Although in essentially the same form as in earlier reports, this table reflects a new and improved method of computing the percentile frequency distribution so as to be more representative of the true distribution.

The first extensive compilation of nonurban manganese determinations to be published was that for quarterly composites of samples collected from 29 stations in 1965 (Table 2-9).[501] These were followed in the next report[510] by 199 individual samples for earlier years, representing eight different locations, of which four were in the 1965 group of 29. The 1965 quarterly averages varied from below the minimum detectable, at several western locations, to 0.080 µg/m³ in Clarion County, Pa.; the averages for the year ranged from 0.0017 µg/m³, for Yellowstone Park, Wyo., to 0.047 µg/m³, for Clarion County, Pa.[501] The maximum for the 199 individual samples reported later for earlier years was 0.08 µg/m³, for the Loquillo Mountains, Puerto Rico.[510]

Any attempt to obtain valid manganese concentration gradients or yearly trends from the available data would appear to be a project in itself, with some question as to whether sufficient adequate comparable data are available and whether a studied analysis is really necessary. It is rather obvious from the published data and our other information that, with the rare exception of a volcanic explosion, high concentrations of manganese in the air depend on man-made sources of emission modified by the previously enumerated qualifying factors of topography and atmospheric conditions. As for yearly trends, it is possible that a study of frequency distributions obtained from the sampling networks, in conjunction with a study of the particular factors and events that could have affected those distributions, would disclose some trends of interest. However, it would require assurance, first, that the sampling and the data on which the calculated distributions depend were such as to justify the conclusions reached. This would call for a careful study of the actual samples used over the entire period under examination, with particular attention to the effect of abnormal high values on the average. It may be found that, to obtain a representative average, ab-

TABLE 2-8 Urban Frequency Distribution of Airborne Manganese[a]

Sampling Site	Year	No. Samples	Manganese Concentration, μg/m³												Arith. mean	Geo. mean	Std. geo. dev.
			Min.	Frequency distribution, percentile										Max.			
				10	20	30	40	50	60	70	80	90					
Alaska Anchorage Site 01	1963	26	0.00	0.00	0.00	0.01	0.02	0.02	0.03	0.04	0.08	0.12	0.18	0.04	0.03	2.62
California Burbank Site 01	1960	26	0.00	0.02	0.03	0.03	0.04	0.04	0.06	0.06	0.07	0.08	0.14	0.05	0.04	1.82
Long Beach Site 01	1965	25	0.00	0.00	0.00	0.01	0.01	0.01	0.01	0.02	0.02	0.03	0.03	0.01	0.01	1.56
Los Angeles Site 01	1960	25	0.00	0.02	0.03	0.03	0.03	0.04	0.04	0.06	0.06	0.08	0.11	0.05	0.04	1.75
Pasadena Site 01	1960	25	0.00	0.01	0.01	0.02	0.04	0.04	0.04	0.05	0.06	0.07	0.38	0.05	0.04	2.41
San Jose Site 01	1963	26	0.00	0.00	0.01	0.02	0.02	0.03	0.03	0.04	0.05	0.06	0.09	0.03	0.03	1.94
Kansas Kansas City Site 01	1962	24	0.01	0.02	0.02	0.02	0.03	0.03	0.04	0.05	0.07	0.07	0.17	0.05	0.03	2.02
Minnesota St. Paul Site 01	1958	23	0.02	0.02	0.03	0.03	0.04	0.05	0.05	0.06	0.06	0.08	0.09	0.05	0.04	1.60
Missouri Kansas City Site 01	1963	24	0.00	0.00	0.01	0.02	0.02	0.03	0.03	0.04	0.06	0.07	0.21	0.04	0.03	2.15

Location	Year	N													
Elizabeth															
Site 01	1962	25	0.00	0.00	0.01	0.02	0.02	0.02	0.03	0.05	0.13	0.15	0.04	0.03	2.28
Glassboro															
Site 01	1965	24	0.00	0.00	0.02	0.02	0.02	0.02	0.02	0.05	0.07	0.09	0.03	0.03	1.97
Jersey City															
Site 01	1963	25	0.00	0.00	0.01	0.02	0.03	0.03	0.03	0.04	0.07	0.10	0.03	0.03	1.94
New York															
Massena															
Site 01	1960	24	0.00	0.00	0.01	0.01	0.01	0.02	0.02	0.03	0.04	0.07	0.02	0.02	1.84
New Rochelle															
Site 01	1960	25	0.01	0.01	0.01	0.02	0.02	0.03	0.03	0.05	0.06	0.11	0.03	0.03	2.03
North Carolina															
Asheville															
Site 01	1962	25	0.00	0.00	0.01	0.01	0.02	0.02	0.02	0.02	0.04	0.05	0.02	0.02	1.68
Oregon															
Medford															
Site 01	1961	22	0.04	0.05	0.07	0.10	0.11	0.12	0.15	0.17	0.20	0.37	0.13	0.12	1.72
Texas															
Ft. Worth															
Site 01	1962	25	0.01	0.01	0.01	0.02	0.02	0.02	0.02	0.04	0.05	0.08	0.13	0.04	0.03	2.18
Houston															
Site 01	1963	22	0.01	0.01	0.01	0.02	0.02	0.02	0.03	0.06	0.11	0.27	0.45	0.08	0.04	3.31
	1965	25	0.00	0.00	0.02	0.03	0.04	0.05	0.08	0.08	0.09	0.57	0.07	0.04	2.55
	1963–1965	47	0.00	0.00	0.01	0.02	0.02	0.03	0.04	0.07	0.08	0.19	0.57	0.08	0.04	2.93
Utah															
Salt Lake City															
Site 01	1958	24	0.00	0.00	0.02	0.02	0.03	0.03	0.04	0.05	0.08	0.10	0.13	0.04	0.03	2.18
	1963	26	0.00	0.01	0.02	0.02	0.02	0.02	0.03	0.04	0.05	0.07	0.12	0.03	0.03	1.94
	1964	24	0.00	0.00	0.01	0.01	0.02	0.02	0.02	0.03	0.03	0.04	0.05	0.02	0.02	1.79
	1958–1964	74	0.00	0.00	0.01	0.02	0.02	0.02	0.03	0.04	0.05	0.07	0.13	0.03	0.03	1.97

[a] Derived from U.S. Environmental Protection Agency, Division of Atmospheric Surveillance.[510]

TABLE 2-9 Nonurban Quarterly and Yearly Averages of Airborne Manganese Concentration of 1965[a]

Station Location	Manganese Concentration, µg/m³				
	1st quar.	2nd quar.	3rd quar.	4th quar.	Yrly. avg.
Arizona					
Grand Canyon Pk.	0.0044	0.0067	0.0071	0.013	0.0078
Maricopa Co.	0.018	0.019	0.018	0.011	0.016
Arkansas					
Montgomery Co.	0.0060	0.0063	0.0057	0.0080	0.0065
California					
Humboldt Co.	0.0036	0.0028	0.0062	0.0033	0.0039
Colorado					
Montezuma Co.	0.0042	0.0045	0.0039	0.0000	0.0031
Indiana					
Parke Co.	0.022	0.024	0.015	0.013	0.018
Iowa					
Delaware Co.	0.0089	0.011	0.018	0.011	0.012
Maine					
Acadia Natl. Pk.	0.0037	0.0079	0.010	0.0080	0.0074
Maryland					
Calvert Co.	0.018	0.010	0.0099	0.014	0.012
Mississippi					
Jackson Co.	0.012	0.0083	0.0085	0.015	0.010
Missouri					
Shannon Co.	0.010	0.017	0.013	0.014	0.013
Montana					
Glacier Natl. Pk.	0.0038	0.0061	0.0075	0.0036	0.0052
Nebraska					
Thomas Co.	0.0042	0.0066	0.0069	0.0031	0.0052
Nevada					
White Pine Co.	0.016	0.0030	0.0023	0.0000	0.0053
New Hampshire					
Coos Co.	0.0030	0.0068	0.0039	0.010	0.0059

Manganese in the Ecosystem

TABLE 2-9 (*Continued*)

	Manganese Concentration, $\mu g/m^3$				
Station Location	1st quar.	2nd quar.	3rd quar.	4th quar.	Yrly. avg.
New Mexico Rio Arriba Co.	0.0040	0.012	0.0041	0.0000	0.0050
New York Cape Vincent	0.015	0.035	0.016	0.050	0.029
North Carolina Cape Hatteras	0.014	0.0061	0.011	0.0077	0.0097
Oklahoma Cherokee Co.	0.0093	0.010	0.020	0.0080	0.011
Oregon Curry Co.	0.0042	0.0017	0.0038	0.0000	0.0024
Pennsylvania Clarion Co.	0.037	0.044	0.027	0.080	0.047
Rhode Island Washington Co.	0.013	0.025	0.0099	0.034	0.020
South Carolina Richland Co.	0.015	0.0087	0.0032	0.014	0.010
South Dakota Black Hills Frst.	0.0000	0.0027	0.0029	0.0076	0.0033
Texas Matagora Co.	0.010	0.0032	0.0051	0.0066	0.0062
Vermont Orange Co.	0.014	0.0064	0.016	0.022	0.014
Virginia Shenandoah Pk.	0.017	0.021	0.016	0.020	0.018
Wisconsin Door Co.	0.0046	0.0090	0.0044	0.0094	0.0068
Wyoming Yellowstone Pk.	0.0027	0.0023	0.0021	0.0000	0.0017

[a] Derived from U. S. Department of Health, Education, and Welfare.[501]

normal values need to be "cut," as is the accepted practice in determining averages for some types of gold ores.

From the data that have been published, it appears that the average manganese content of urban air in the United States since 1953 has been approximately 0.10 μg/m^3. As one leaves the urban areas, manganese content of the air drops to below the limits of detectability (that is, essentially none) at favored locations having the right atmospheric conditions at the time of sampling. In fact, as shown by the foregoing report summaries, this has even been achieved in at least one so-called urban area.

CYCLING OF MANGANESE IN THE BIOSPHERE

Precipitation and Water

On the basis of atmospheric precipitation sampling for six metals at 32 widespread stations in the United States, studies by the Laboratory of Atmospheric Sciences, National Center for Atmospheric Research, Boulder, Colo., concluded that manganese in atmospheric precipitation was derived primarily from human activity. The samples were collected from September 1966 through January 1967 and had an average manganese concentration of 0.012 ppm in the precipitate. In grams deposited on 1 hectare in a month—a figure directly proportional to both manganese concentration and quantity of rainfall, snow, and sleet—monthly averages ranged from zero at Mauna Loa Observatory, Hawaii; Amarillo, Tex.; and Tampa International Airport, Fla., to 54 at Caribou, Maine; 31 and 26 at two Chicago airports; and 23 at Sault St. Marie, Mich.[297] The high Maine sample seems to be an anomaly with respect to direct association with human activity, but averages at that station are also high for four of the other metal ions sampled. The monthly manganese averages for the other 31 stations directly support the study's conclusion of a relation to human activity. It is possible that air currents or some sort of meteorologic zoning can justify an indirect relation in the case of Caribou.

From a 1962 U.S. Geological Survey study[144] of public water supplies of the 100 largest U.S. cities, the average manganese concentration of the untreated surface waters sampled was calculated to be 0.070 ppm.[297] In reporting on the results of the study, Durfor and Becker[144 (p.16)] state that the concentration of manganese in natural water is generally 0.20 ppm or less, but that ground water and acid mine water may contain more than 10 ppm. The manganese content of reservoir water that has "turned over" may be more than 150 ppm. They give the maximal manganese concentration in public water supplies of the 100 cities as 0.60 ppm for untreated water and 2.5 ppm for treated water; minima are 0.00 ppm for both treated and untreated water. Using chemical analyses, they found[144 (p.68)] that 95% of the treated water supplies had manganese concentrations of less than 0.10 ppm; by spectrographic analyses, 97% had less than 100 µg/liter (0.10 ppm). The maximum, median, and minimum determined by chemical analysis were 2.50, 0.00, and 0.00 ppm, respectively; and by spectrographic analysis, 1,110 µg/liter, 5.0 µg/liter, and "not detected."[144 (p.78)]

Kopp and Kroner[290] summarized trace-element data collected by local, state, and federal government agencies and private industry in the 5-year period from October 1, 1962, through September 30, 1967, under the Federal Water Pollution Control Administration's water quality surveillance program. They state that the manganese concentration in most natural waters is less than 20 µg/liter (0.02 ppm), but that manganese concentrations above 1 mg/liter (1 ppm) may result where the manganese contained in minerals is dissolved under reducing conditions or under the influence of particular bacteria. Higher values were obtained where mining or industrial wastes polluted the waters. The highest manganese concentrations obtained in the 5-year program were 3.2 mg/liter, in the Allegheny River, and 2.2 mg/liter, in the Monongahela River, both at Pittsburgh. The mean manganese concentrations at these two stations were 0.5 and 0.6 mg/liter, respectively; the minima were 1 and 6.6 µg/liter. By the time the first downstream station on the Ohio was reached at Toronto, Ohio, 19 miles from the Pennsylvania state line, manganese content had dropped to a high of 376, a mean of 68, and a low of 2.7 µg/liter. At Cairo, Illinois, the maximal, mean, and minimal manganese concentrations of the Ohio River were 8.1, 2.9, and 0.4 µg/liter. In other watersheds, the Cuyahoga River at Cleveland, Ohio, reported high values: a maximum of 900, a mean of 244, and a minimum of 4.5 µg/liter.

Of a total of 1,577 samples collected over the conterminous United States and Alaska, manganese was detected in 810, for a frequency of 51.4%. Divided into 16 drainage basins, the percent frequency of detection of various trace metals is shown in Table 2-10; the mean concentrations detected are shown in Table 2-11. The number of samples exceeding the recommended manganese limit of 0.05 mg/liter for public water supplies is shown in Tables 2-12 and 2-13. Table 2-14 compares data on suspended concentrations for 228 samples with all the data

TABLE 2-10 Detection of Trace Metals, Percent Frequency by Drainage Basin[a]

Metal	Northeast	North Atlantic	Southeast	Tennessee River	Ohio River	Lake Erie	Upper Mississippi	Western Great Lakes	Missouri River	Southwest–Lower Mississippi	Colorado River	Western Gulf	Pacific Northwest	California	Great Basin	Alaska
Zn	95.6	94.7	96.7	73.5	81.8	87.3	70.5	90.9	53.0	62.6	45	48.9	90.1	72.4	68.4	83.3
Cd	4.4	7.6	1.1	0	2.9	8.5	1.8	3.0	0	0	1	2.1	2.5	0	5.3	0
As	5.5	7.6	8.8	1.5	8.3	4.3	7.1	4.5	2.0	1.3	2	4.3	8.6	0	5.3	16.7
B	100	99.4	94.5	100	99.2	100	98.2	100	97.3	99.4	98	97.9	93.8	96.5	94.7	100
P	60.4	55.6	51.6	39.7	48.8	53.2	77.7	48.5	39.5	30.3	22	23.4	52.5	62.1	26.3	55.6
Fe	87.9	78.9	98.9	83.8	68.6	66.0	80.4	78.8	66.0	80.0	59	70.2	80.2	93.1	73.7	94.4
Mo	13.2	32.7	18.7	38.2	28.1	27.7	68.8	51.5	32.0	20.0	37	10.6	38.9	37.9	57.9	27.8
Mn	50.5	48.5	72.5	57.4	58.7	61.7	61.6	56.1	40.8	34.8	39	38.3	51.2	44.8	57.9	61.1
Al	67.0	30.4	71.4	47.1	21.5	21.3	19.6	42.4	11.6	19.4	14	21.3	30.2	17.2	15.8	27.8
Be	1.1	23.4	1.1	1.5	14.1	6.4	0	1.5	2.0	0	0	0	0.6	0	0	0
Cu	97.8	92.3	95.6	91.2	72.3	50.4	80.4	97.0	49.7	63.9	33	40.4	87.0	69.0	73.7	94.4
Ag	14.3	5.3	5.5	0	5.4	6.4	5.4	9.1	4.1	4.5	18	4.3	8.6	0	5.3	5.6
Ni	22.0	28.1	20.9	8.8	25.2	53.2	15.2	9.1	2.0	9.7	8	2.1	10.5	13.8	15.8	11.1
Co	2.2	1.2	1.1	0	9.9	4.3	0.9	3.0	0.7	0.7	2	0	3.7	0	0	0
Pb	36.3	22.2	12.1	35.3	18.6	27.7	24.1	40.9	3.4	9.7	15	2.1	22.8	6.9	21.1	38.9
Cr	56.0	21.1	40.7	47.1	23.6	23.4	17.9	28.8	4.8	20.0	17	6.4	32.7	20.7	10.5	22.2
V	4.4	3.5	1.1	0	3.7	4.3	0.9	0	1.4	3.9	9	2.1	6.2	6.9	0	5.6
Ba	100	100	100	100	100	100	100	100	98.0	100	100	100	100	100	94.7	100
Sr	100	100	98.9	100	100	100	100	100	99.3	98.7	100	100	99.4	96.5	94.7	100

[a] Derived from Kopp and Kroner.[290]

TABLE 2-11 Mean Concentrations of Trace Metals, by Drainage Basin, μg/liter[a]

Metal	Northeast	North Atlantic	Southeast	Tennessee River	Ohio River	Lake Erie	Upper Mississippi	Western Great Lakes	Missouri River	Southwest–Lower Mississippi	Colorado River	Western Gulf	Pacific Northwest	California	Great Basin	Alaska
Zn	96	49	52	28	81	205	45	24	39	85	51	92	40	16	44	28
Cd	5	3	5	—	7	50	6	5	—	—	2	10	5	—	1	—
As	34	47	35	50	66	308	69	37	123	91	53	22	68	143	20	34
B	32	42	29	24	67	210	105	19	154	131	179	289	30	83	84	28
P	44	48	43	42	130	153	243	31	353	81	121	173	47	46	37	40
Fe	51	19	120	37	28	35	35	22	37	69	40	173	32	45	70	25
Mo	25	33	15	25	70	68	88	28	83	95	130	24	30	—	145	17
Mn	3.5	2.7	2.8	3.7	232	138	9.8	2.3	13.8	9.0	12	10	2.8	2.8	7.8	18
Al	28	22	117	30	141	56	18	17	213	68	50	333	30	63	15	11
Be	0.02	0.12	0.05	0.16	0.28	0.17	—	0.05	0.23	—	—	—	0.02	—	—	—
Cu	15	17	14	11	23	11	14	7	17	19	10	11	9	12	12	9
Ag	1.9	0.9	0.4	—	2.1	5.3	3.4	1.4	1.2	4.3	5.8	3.5	0.9	—	0.3	1.1
Ni	8	8	4	4	31	56	15	10	5	17	12	3	10	10	4	5
Co	14	9	1	—	19	33	18	11	8	36	11	—	8	—	—	—
Pb	17	14	8	17	30	39	33	14	39	37	32	4	15	4	18	12
Cr	14	6	4	6	7	12	7	6	17	16	16	25	6	15	4	9
V	9	12	10	—	22	54	20	—	171	25	105	9	13	30	—	32
Ba	21	25	26	25	43	42	39	15	63	90	60	67	27	42	41	17
Sr	76	62	26	47	130	260	105	44	342	540	697	652	68	153	152	81

[a] Derived from Kopp and Kroner.[290]

TABLE 2-12 Violations of Water-Quality Criteria[a]

Metal	Limit, μg/liter	No. Violations	Mean of Values That Exceeded the Limits, μg/liter
Cadmium	10	6	39
Arsenic	50	41	91
Iron	300	25	>676
Manganese	50	74	>586
Lead	50	27	71
Chromium (hexavalent)	50	4	94
Zinc	5,000	0	–
Copper	1,000	0	–
Silver	50	0	–
Barium	1,000	0	–

[a] Derived from Kopp and Kroner.[290]

obtained by the 1,577 samples for metals in solution. Table 2-15 compares the dissolved and suspended manganese contents for several selected stations along five eastern rivers.

In one of the first extensive applications of spectrographic analysis to the determination of trace elements in water, Braidech and Emery[72] reported on the water supplies of 24 municipal water plants distributed

TABLE 2-13 Violations of Water-Quality Criteria, by Drainage Basin[a]

Basin	Total No.	Cd	As	Fe	Mn	Pb	Cr
Northeast	6	1	1	2	0	0	2
North Atlantic	7	0	6	0	0	1	0
Southeast	12	0	2	10	0	0	0
Tennessee River	2	0	1	1	0	0	0
Ohio River	81	2	15	0	58	6	0
Lake Erie	18	2	0	1	12	3	0
Upper Mississippi	11	0	3	1	1	6	0
Western Great Lakes	3	0	1	0	0	2	0
Missouri River	6	0	3	1	1	1	0
Southwest–lower Mississippi	10	0	1	3	1	4	1
Colorado River	5	0	1	0	0	3	1
Western Gulf	6	0	0	6	0	0	0
Pacific Northwest	9	1	7	0	0	1	0
California	0	0	0	0	0	0	0
Great Basin	0	0	0	0	0	0	0
Alaska	1	0	0	0	1	0	0

[a] Derived from Kopp and Kroner.[290]

TABLE 2-14 Comparison of Suspended and Dissolved Trace Metals in Surface Waters[a]

Metal	Suspended Metals in 228 Samples[b]			Dissolved Metals in 1,577 Samples		
	No.	%	Mean Concentration, µg/liter	No.	%	Mean Concentration, µg/liter
Zn	146	64	62	1,207	76.5	64
Cd	0	0	–	40	3	9.5
As	–	–	–	–	–	–
B	52	23	44	1,546	98	101
P	–	–	–	–	–	–
Fe	228	100	3,000	1,192	76	52
Mo	–	–	–	–	–	–
Mn	212	93	105	810	51	58
Al	221	97	3,860	456	31	74
Be	40	18	0.34	85	5	0.19
Cu	141	62	26	1,173	74	15
Ag	0	0	–	104	7	2.6
Ni	7	3	29	256	16	19
Co	0	0	–	44	3	17
Pb	5	2	120	305	19	23
Cr	18	8	30	386	25	9.7
V	0	0	–	54	3	40
Ba	216	95	38	1,568	99	43
Sr	22	10	58	1,571	100	217

[a] Derived from Kopp and Kroner.[290]
[b] Current[290] laboratory practice is to analyze both dissolved and suspended fractions. To date, this has been done for 228 samples. This table compares suspended concentrations observed in the 228 samples with the complete body of data on dissolved material.

over the United States. Manganese content of monthly composites made from daily samples taken for February and March ranged from 0.000 ppm, for the Mississippi River at both New Orleans and St. Louis, the Missouri River at Bismarck, and Lake Erie at Cleveland, to 0.3 ppm, for Indianapolis (White River watershed), Pittsburgh (Allegheny River), and Philadelphia (Schuylkill and Delaware rivers).

The recommended upper limit for manganese, established in a Drinking Water Standard by the U.S. Public Health Service in 1962, is 0.05 mg/liter (0.05 ppm).[297,503,508] This was determined primarily for aesthetic and economic reasons, no harmful concentration being known at that time for ingestion of manganese. Difficulties encountered in attempting to remove manganese to residual concentrations much below 0.05 mg/liter and in measuring such concentrations influenced the decision to set this as the limit.[503] Nevertheless, the American Water Works Association has set a goal of 0.01 mg/liter, again for aesthetic reasons, as providing the ideal that might be generally attainable by

TABLE 2-15 Comparison of Suspended and Dissolved Manganese in River Waters[a]

Sampling Point	Dissolved Manganese			Suspended Manganese		
	Average of Observed Concentrations, µg/liter	Range of Observed Concentrations, µg/liter	Occurrence, %	Average of Observed Concentrations, µg/liter	Range of Observed Concentrations, µg/liter	Occurrence, %
Delaware River						
at Martins Creek, Pa.	2.9	1.3–4	40	18	3–51	93
at Trenton, N. J.	3.2	1.1–6.2	27	57	10–107	100
at Philadelphia, Pa.	4.2	3.2–5.8	31	68	29–120	100
Allegheny River						
at Pittsburgh, Pa.	>1,000	74–3,230	60	205	18–1,500	100
Monongahela River						
at Pittsburgh, Pa.	607	6.6–2,150	93	73	2–442	87
Ohio River						
below Addison, Ohio	57	7.8–180	51	238	140–400	100
Kanawha River						
at Winfield Dam, W. Va.	44	4.6–115	87	35	2.5–84	100

[a] Derived from Kopp and Kroner.[290]

Manganese in the Ecosystem

correct application of known water treatment methods.[413] The tentative limits suggested for irrigation waters by the Committee on Water Quality Criteria, Federal Water Pollution Control Administration, are 2.0 mg/liter for continuous use on all soils and 20.0 mg/liter for short-term use on fine-textured soil.[290,509]

A 1969 Public Health Service (PHS) study[508] of 969 community water supply systems found that the average delivered water from 90 of the systems exceeded the PHS recommended manganese limit. A maximal manganese concentration of 1.32 mg/liter was found when 2,595 individual samples collected "at a consumer's tap" were analyzed. Of these 2,595 samples, 211 exceeded the recommended manganese limit.

Samples, obtained in April–June 1968, of water delivered by two City of Chicago water treatment plants did not exceed 0.002 mg/liter on analysis. Somewhat higher concentrations obtained at consumer's taps supplied by these plants fault the distribution system or are associated with the consumer's facilities. None exceeded 0.025 mg/liter.[336]

An unusual case of manganese contamination of well water is reported in the Japanese literature. In August–October 1939, more than 300 discarded manganese dry cells were buried to a depth of about 2 m around a well, and another 100 cells were buried approximately 9 m away. Analysis of the well water on December 18 showed it to contain 14.34 mg of manganese tetroxide per liter. All the cells were removed from both locations. On January 18, the well water was analyzed at 8.75 mg/liter, water from a well approximately 11 m away was analyzed at 10.65 mg/liter, water from a well approximately 46 m from the first was analyzed at 7.63 mg/liter, and water from two other wells approximately 15 and 27 m from the first, but in the opposite direction, showed no manganese tetroxide. Analysis of the first well on February 5 showed 4.20 mg/liter.[275]

Soils and Biota

Manganese in soils depends primarily on the manganese content of the parent rocks, which may have an *in situ* relation or be a considerable distance removed. Physical, chemical, and biologic actions—as represented or affected, for example, by weathering, drainage, or transportation—have their effect on the manganese content, with the state of the manganese and the pH of the soil being important considerations. The natural occurrences of manganese in soils—i.e., those occurrences essentially unaffected by human activities—have already been discussed.

"Reports of contamination of soils with Mn are few. Mine seepage, fly ash, and impure phosphate fertilizers have been held responsible,"[295] contamination being defined as the introduction of undesirable foreign substances into the plow layer. Five possible sources of heavy-metal contamination are identified by Lagerwerff as aerosols, pesticides, limestone and phosphate fertilizers, manures and sewage sludges, and mine wastes. Typical manganese contents of some of these possible contaminants are as follows:[295]

rock phosphate	85 mg/kg dry weight[96]
limestone	850 mg/kg dry weight[434]
superphosphate	51 mg/kg dry weight[89]
sewage sludge	240 mg/kg dry weight[16,96]
spoil-bank effluent	1,800 mg/liter[330]

Manganese contents of representative samples of sewage sludge obtained in 1952 from eight different locations in the United States, reported as used for soil improvement, ranged from 60 to 440 ppm. Two other sludges, one from an agricultural research center and the other from a leather-goods manufacturing plant, had contents of 790 and 760 ppm, respectively.[16] Of more than 280 different pesticides listed in 1964, manganese was present in only 12.[295,341]

Samples of washed pasture grass were collected in 1969 along a traverse perpendicular to a major highway at Denver, Colo., at approximate distances of 1.5, 3, 15, 30, 61, 152, 229, and 305 m from the highway. Contrary to the experience for lead and some of the other metals for which analyses were made, there was no decrease in manganese content as the distance from the highway increased.[85]

Although there has been some recent sampling of soils with the objective of determining the relation of metal content to motor traffic, manganese does not appear to have been of interest in most instances. An exception has been a recent program of the Environmental Protection Agency. A small number of samples taken from approximately 5 cm below the surface in the front and back yards of houses in three urban residential areas showed no consistent traffic relation with manganese. This, again, was contrary to the experience for lead and some other metals for which analyses were made.[404]

As stated earlier, all or nearly all organisms contain manganese.[226,309,442] Marine organisms are capable of concentrating manganese in their bodies to many times above the concentration in seawater; manganese enrichment factors reported for shellfish, compared with their marine environment, range from 1,100 to 13,500,[410] and for algae, as much as 60,000 (from 0.002 mg/kg in seawater to 120 mg/kg in the algae).[349,527] Schroeder and co-workers,[442,443] using data from Vinogradov[516] and

Lounamaa and taking seawater at 0.001 ppm, report concentrations of manganese in the bodies of marine plants and fish of up to 100,000 and 100, respectively, times the concentration in seawater. They give a concentration factor of 10 for terrestrial mammals, and three or four for man. According to emission spectrography, all human tissues contain manganese,[443] the total content for a 70-kg adult being approximately 12 mg.[442] The manganese contents of various foods[442,443] are shown in Table 2-16.

TABLE 2-16 Manganese in Food[a]

Item	Manganese Concentration (wet wt)	
	µg/g	µg/100 calories
Grains and cereals		
Wheat, whole, seed	11.32	340
Wheat, flour, all-purpose	5.20	150
Bread, white	1.78	67
Bread, whole wheat	1.43	38
Buckwheat, whole, seed	13.11	391
Rye, whole, seed	13.29	397
Oats, seed	6.46	160
Oatmeal	2.72	67
Barley, seed	17.80	494
Corn, whole	1.31	37
Corn, meal	2.05	59
Rice, American, brown	1.42	39
Rice, Madagascar	1.17	33
Rice, Japanese, unpolished	2.08	58
Rice, Japanese, polished	1.53	43
Millet, seed	2.04	57
Macaroni, dry	10.56	293
Grapenuts	30.76	854
Dairy products		
Milk, whole, bulk	0.19	29
Milk, direct from udder	0.14	21
Milk, dry, skimmed	0.0	0
Butter	0.96	12
Cheese, Swiss, domestic	1.32	33
Cheese, Gorgonzola, domestic	1.88	48
Eggs, whole	0.53	33
Egg whites	0.43	116
Egg yolks	0.88	25

TABLE 2-16 (*Continued*)

Item	Manganese Concentration (wet wt)	
	µg/g	µg/100 calories
Meat and poultry		
Beef, roasting	0.05	3
Beef, brisket	0.03	1
Beef liver	0.16	11
Beef kidney	0.0	0
Pork loin	0.0	0
Pork chops, lean	0.34	12
Pork liver	0.09	6
Pork kidney	0.75	63
Lamb chops, lean	0.34	18
Lamb kidney	0.30	30
Chicken breast	0.21	11
Fish and seafood		
Oysters, fresh, frozen	0.06	12
Clams, fresh, frozen	0.0	0
Scallops, fresh	0.11	11
Shrimp, fresh, frozen	0.02	2
Cod steak	0.02	3
Halibut steak	0.12	9
Herring, smoked	0.04	2
Sardines, canned, Morocco	0.0	0
Anchovies, canned, Portugal	0.10	5
Fruit		
Orange, Florida	0.35	100
Tangerine, Florida	0.62	180
Peach	1.02	275
Pear, Anjou	0.33	83
Apple, MacIntosh	0.31	70
Raisins, packaged	4.68	190
Apricot, canned	0.20	33
Cantaloupe	0.26	108
Nuts (edible part only)		
Pecans	35.09	585
Brazil	27.80	432
Almonds	24.69	412
Walnuts	7.52	137
Coconut, fresh, Virgin Islands	0.38	10
Peanuts, salted	6.91	113
Vegetables		
Beans, green, canned	0.24	70
Peas, fresh	0.64	100
Peas, split green, dry	12.74	463

TABLE 2-16 (*Continued*)

Item	Manganese Concentration (wet wt)	
	µg/g	µg/100 calories
Carrots, canned	1.56	780
Beets, fresh	0.41	145
Beet greens, fresh	4.26	3873
Beets, canned	1.34	304
Turnip greens, fresh	4.71	4282
Spinach, canned	1.42	546
Spinach, fresh	7.77	2990
Cabbage, fresh	1.13	435
Broccoli, fresh	1.54	1000
Brussels sprouts, fresh	3.45	2156
Asparagus, fresh	0.32	178
Asparagus tips, canned	0.27	150
Cucumber, fresh	0.14	156
Rhubarb, fresh	4.57	7617
Tomatoes, canned	0.30	214
Tomato soup, canned	0.32	230
Condiments and beverages		
Salt, table	0.0	—
Paper, black	47.48	—
Paprika	4.86	—
Cloves	262.86	—
Ginger, ground	87.33	—
Thyme, ground	82.65	—
Bay leaf, dried (laurel)	67.19	—
Garlic, fresh	0.95	—
Garlic powder	0.45	—
Chili powder	0.62	—
Caraway seeds, whole	2.02	—
Mustard, dry	1.26	—
Vinegar, cider	2.54	—
Coffee, ground	20.65	—
Coffee, infusion	0.85	—
Tea, leaves	275.58	—
Tea, infusion	6.9	—
Fats and oils		
Cod liver oil, Norway	4.95	55
Sunflower	1.47	16
Safflower	0.0	0
Cottonseed	2.52	28
Corn	1.00	11
Peanut, pressed	1.00	11
Linseed, pressed	4.20	47
Lard, refined	0.98	10

[a] Derived from Schroeder et al.[443]

SUMMARY

Manganese is one of the more abundant metals of the earth's crust. It is widely distributed in rocks, soils, fresh waters, seawaters, and sediments, and it is present in all or nearly all organisms. Its concentration is variable within each of these categories, often measured in parts per million, or in parts per billion in the case of waters. In rocks and soils, it ranges up to ores containing 50% manganese; sea-floor deposits have reported individual samples that high in manganese.

There are numerous valence states for manganese, with the divalent form giving the most stable salts, and the tetravalent, the most stable oxide. At least 100 minerals contain manganese as an essential element, and perhaps 200 others contain it as an accessory element. Its ores have wide distribution over the tropical, subtropical, and warmer temperate zones of the earth.

Manganese occurs in nature in constantly changing complex relations involving physical, chemical, and biologic activity, with the oxides being the dominant form, although carbonates and silicates occur commonly as minerals. Any study of the data must cope not only with the actual variations in content but also with variations resulting from different methods of sampling and analysis, which are of uncertain reliability and interpretation. This is particularly true with respect to water, with its complexity, low concentrations, and problems of sampling and analysis. Two samples of water taken at the same place at different times cannot necessarily be expected to show the same composition.

Man's principal use of manganese is in the production of steels and cast irons to nullify the harmful effects of sulfur, and approximately 90% (or more) of its consumption is by the iron and steel industry for this purpose or as an alloying agent. Steel cannot be made in quantity without it. It is commonly added in the form of ferromanganese or silicomanganese and less commonly as manganese metal or spiegeleisen. Manganese is important to the production of aluminum, magnesium, and some copper alloys. High-carbon ferromanganese and spiegeleisen are made in either blast or electric furnaces; the other manganese ferroalloys are made, with few exceptions, in electric furnaces; and manganese metal is most commonly the product of electrolysis.

Dioxide ores, either as particular ores or as a refined chemical, are essential constituents of the common dry-cell battery. The same or other ores have a variety of chemical and miscellaneous uses.

In considering manganese emissions to the atmosphere as a result of various human activities, one does not find much in the way of good recent data. The data that are available must be used with care, taking into account various qualifying factors, such as geographic location, atmospheric conditions, degree of control, and nature of the sampling and analysis.

In the past, the blast and the electric manganese ferroalloy furnaces have been the worst offenders. Examples of efficient cleaning systems are reported, but there does not appear to be any documentation of their extent of application or of the degree of their overall efficiency today. Although total volumes on an uncontrolled basis are much greater, emissions from steel furnaces and pig-iron blast furnaces have much lower manganese concentrations, are more easily controlled, and apparently are being efficiently controlled for the most part.

Emission of dusts arising from the handling of manganese ores and other manganiferous materials (including recovered dusts) and emission of fumes and dusts from foundries and miscellaneous uses can be controlled and probably are to be considered essentially local problems where they are not controlled. There appears to be little or no published information on such emission to the general atmosphere.

Manganese emission from the burning of coals and fuel oils apparently does not normally present a serious problem, particularly when one considers the cleaning practices in use.

The U.S. Public Health Service initiated an air sampling program in 1953, which has grown into the National Air Surveillance Networks of the Environmental Protection Agency. Analyses for manganese that have been obtained in the program range from below the minimum detectable, for several nonurban western locations, to more than 10 $\mu g/m^3$, for some industrial urban sampling locations. The average manganese concentration for urban air over the United States from 1953 through 1965, the last year for which sample results have been published, was approximately 0.10 $\mu g/m^3$.

From a study of atmospheric precipitation sampling for six metals at 32 stations distributed over the United States, it was concluded that manganese in atmospheric precipitation was primarily the result of man's activities. The average manganese concentration was 0.012 ppm in the precipitate. From a 1962 U.S. Geological Survey study of public water supplies of the 100 largest U.S. cities, it was calculated that the average manganese concentration of untreated surface waters was 0.070 ppm. This compares with the Public Health Service Drinking Water Standard's recommended upper limit of 0.05 ppm, which was established in 1962 primarily for aesthetic and economic reasons, in view of the fact that no concentration was known to be dangerous to health.

This and other surveys disclosed mining and industrial pollution of waters, brought out the fact that the introduced manganese concentration was soon dissipated or diluted in rivers, and showed great differences in manganese concentration between samples taken at the same station at different times.

Mine effluents, fly ash, and impure fertilizers and soil additives are the chief sources of contamination of soils by manganese, but there are few reports of such contamination. The manganese content of soils depends primarily on the manganese content of the parent rocks as modified by physical, chemical, and biologic factors, such as weathering, drainage, and transportation.

Manganese in the ecosystem has well-established lines of movement from rocks to soils to plants to animals, from soils to water to organisms, and back to water and soils. Marine organisms can concentrate manganese in their bodies to many times the concentration in seawater, and man retains manganese at a concentration three or four times higher than that in his food.

There are, in addition, less-apparent movements of manganese among the components of the ecosystem—e.g., from atmosphere to water to soils, from water to sediments and vice versa, from soils to atmosphere with or without man's assistance, and even from soils to rocks (if one thinks in terms of geologic time).

3

Manganese and Plants

SOIL CHARACTERISTICS THAT AFFECT MANGANESE AVAILABILITY

The availability of manganese to plants is affected by numerous soil characteristics, including the concentration of total or easily reducible manganese in soil parent materials, concentrations of other cations and total salts, pH, cation-exchange capacity, drainage, organic matter content, temperature, compaction, and microbial activity.[5,259,272,293,365] Small changes or inherent differences in these soil characteristics can determine whether the soil content of available manganese will be deficient, adequate, or toxic for a given crop.

Plants apparently absorb manganese primarily in the divalent state.[5] Lowering the soil pH or reducing soil aeration by flooding or compaction favors the reduction of manganese to this form and thereby increases its solubility and availability to plants.[101,206,365] Heavy fertilization of acid soils without liming—particularly with materials containing chlorides, nitrates, or sulfates—may also increase manganese solubility and availability.[92,313,394,536] Tables 3-1 and 3-2 show that fertilization without lime greatly increased the uptake of manganese by tobacco. Many of these salt effects are associated with decreased soil pH, but specific anion effects have also been postulated.[92,536]

TABLE 3-1 Effects of Soil Fertility and pH on Manganese Concentrations of Upper and Lower Tobacco Leaves (Averages of Three Varieties)[a]

Soil Fertility	Soil pH	Manganese Concentration in Leaves, ppm (dry wt)	
		Upper[b]	Lower[c]
Low	5.8	56	124
Low	6.4	30	52
High	5.4	153	369
High	6.8	33	61

[a] Derived from Lockman.[313]
[b] Fourth leaf from top.
[c] Fourth leaf from bottom.

Phosphate fertilization may increase or decrease manganese availability. For example, Messing[350] found that the addition of superphosphate to soils with a pH below 5.0 raised the pH and decreased the concentrations of water-soluble and exchangeable manganese; however, when added to soils with a pH above 5.5, superphosphate decreased the pH and increased manganese solubility. The addition of organic matter to soils generally reduces the availability of both native and applied manganese.[177,394,436]

The availability of soil manganese is closely related to the activities of microorganisms that alter pH and oxidation-reduction potentials.[78] Air drying, steam sterilization, and treatment of soils with formalin or other fumigants may increase manganese solubility to a toxic degree in high-manganese soils.[184,212,454] Some of these effects may be due to a

TABLE 3-2 Effects of Soil Fertility and pH on Manganese Concentrations of Upper Leaves of Three Tobacco Varieties[a]

Soil Fertility	Soil pH	Manganese Concentration in Upper Leaves, ppm (dry wt)		
		Burley 21	Delcrest	N.C. 95
Low	5.8	45	66	57
Low	6.4	10[b]	53	28
High	5.4	106	295	59
High	6.8	10[b]	41	49

[a] Derived from Lockman.[313]
[b] Deficient at soil pH above 6.3.

reduction in the populations of manganese-oxidizing organisms.[541] However, Cheng and Ouellette[93] observed that steaming a soil that was high in iron did not increase manganese availability. Adding organic matter to soils can reduce manganese availability by increasing populations of such manganese-oxidizing microorganisms and by forming organic manganese complexes of reduced solubility or availability.[177,490] Geering et al.[190] found that 84–99% of the total manganese in solutions displaced from soils of several areas was in chelated form. Under some conditions of pH and aeration, the addition of organic compounds to soil can increase the chemical reduction of manganese and its uptake by plants.

Manganese deficiencies generally occur in organic soils and neutral or alkaline mineral soils.[51,293] Hanna and Hutcheson[213] reported that manganese deficiency is most common at a pH between 6.5 and 8.0. Some sandy soils may lack sufficient cation-exchange sites (provided by clay or organic matter) to keep manganese from leaching out of the profile and causing a deficiency of total manganese. For example, Gladstones and Drover[198] found that manganese absorption by four lupine species was directly proportional to the clay content of the soils on which they were grown. Sparr[463] has reviewed the manganese fertilization needs for different crops and soil regions of the United States.

TESTING SOIL FOR MANGANESE DEFICIENCY AND TOXICITY

Deficiency

Various soil extractants have been used to predict manganese deficiency for different crops and soils. Total manganese content is useful in determining potential manganese availability in a given soil situation but does not necessarily reflect the actual solubility of manganese or its availability to a plant. Toth and Romney[494] found that the manganese-supplying powers of 20 important New Jersey soils for alfalfa were closely related to their exchangeable and easily reducible manganese supplies. For crops in general, adequate concentrations of exchangeable manganese lie between 0.2 and 3 ppm,[220,223,449] and concentrations of easily reducible manganese between 9 and 60 ppm.[303,450] According to Leeper,[303] heavy liming of soils containing less than about 6–15 ppm of easily reducible manganese may cause manganese deficiency. Hoff and Mederski[235] concluded that manganese deficiency in soybeans may be expected in Ohio soils that contain less manganese than 20 ppm when extracted with 0.1 N H_3PO_4 or 3 N $NH_4H_2PO_4$ or 62.5 ppm when extracted with alcoholic hydroquinone. Anderson and Boswell[17] found that cotton yields were increased by banded manganese fertilizer in Georgia soils that contained water-soluble manganese at up to 2.3 ppm; however, yields were either unaffected or decreased by manganese applications to soils that had higher concentrations of water-soluble manganese. Yield responses were generally obtained in soils that contained water-soluble manganese at less than 1.0 ppm. Soil extraction with DTPA (diethylenetriaminepentaacetic acid) has been used to diagnose manganese deficiency in soils in Colorado.[172]

Hoyt and Nyborg[243] extracted manganese from 40 acid Canadian soils (pH, 4.15–5.63) and correlated the concentrations with manganese uptake by barley, rape, and alfalfa. A 16-h extraction with 0.01 M calcium chloride gave by far the best estimate of available manganese, but extraction with 0.1 N acetic acid also gave a fairly good index of manganese uptake by plants. Only fair to poor estimates were obtained with the other five extractants—0.1 N phosphoric acid and 0.002 N hydrochloric acid for soluble manganese, 1 N potassium chloride and 1 N ammonium acetate (pH, 3.0) for exchangeable manganese, and 0.2% hydroquinone in 1 N ammonium acetate (pH, 7.0) for easily reducible manganese. Total manganese led to the poorest estimate of manganese availability in soils. Manganese concentrations in plants were more closely related to soil pH than to the total soil manganese concentration or cation-exchange capacity.

To be useful, any soil extraction procedure for predicting manganese availability must be applied within a well-defined set of circumstances regarding soil, climate, and plant species or variety. For additional details on soil tests for manganese and other micronutrients, see Adams[4] and Jones.[264]

Toxicity

In moderately well-drained soils, manganese toxicity is generally found only if the soil pH is below about 5.5;[5] however, in flooded soils, the reducing conditions produced can result in temporary toxic concentrations of divalent manganese at a pH approaching 7.0.[206] This reduction process is favored by higher soil temperatures.[94] Potential manganese toxicity may also result from the application of manganese-containing fertilizers. Reuther et al.[416] reported a twofold to threefold increase in the total manganese concentrations of Florida surface soils (Valencia orange groves) during a 7-year period.

Many extraction procedures have been proposed to predict toxic concentrations of manganese in acid soils. Morris[361] used water-soluble manganese (a soil:water ratio of 1:2) as an index of toxicity in 25 naturally acid soils. Concentrations of 0.5, 1.0, and 2.1 ppm (dry-soil basis) were obtained for soils having pH's of above 5.4, 5.2–5.4, and below 5.2, respectively. Adams and Wear[6] found that manganese toxicity in cotton was more closely correlated with water-soluble than with exchangeable or reducible manganese content in some soils of Alabama. Parker et al.[394] reported that manganese toxicity in soybeans was associated with a water-soluble manganese content of 2.5 ppm in an acid soil during a prolonged wet period. In nutrient cultures, a manganese content as low as 0.5 ppm (in a standard Hoagland solution) was toxic to Atlas 46 barley.[543]

Alfalfa grown in greenhouse pots of several acid soils from the southeastern United States showed manganese toxicity when the air-dried soil content of exchangeable manganese (1 N ammonium acetate at a pH of 7.0) was about 50 ppm or above.[177] The highest alfalfa yields were obtained by liming the soils to a pH of about 5.4, which reduced the concentration of exchangeable manganese to about 20 ppm. However, exchangeable manganese is not always a reliable indicator of toxicity. For example, one soil having 136 ppm was not toxic to alfalfa, because a higher content of exchangeable calcium reduced the manganese uptake. Conversely, in later studies, an acid soil that contained exchangeable manganese at only 7.8 ppm, but very low exchangeable calcium, produced manganese toxicity in Pima S-2 cotton.[178]

MANGANESE UPTAKE AND UTILIZATION

Functions of Manganese

The divalent manganese ion activates many enzyme reactions involved in carbohydrate breakdown and in the metabolism of organic acids, nitrogen, and phosphorus;[370] it is also involved in photosynthesis.[15,210] There is evidence that manganese is firmly bound to the lamellae of chloroplasts[393] and that it has a role in maintaining chloroplast structure.[408] Manganese deficiency was characterized by a disorganization of the intergrana membrane connections. Manganese may also play a part in stabilizing ribosome structure.[323] For a discussion of how manganese and other metals may participate in enzyme reactions, see Wallace.[528 (p.257)]

Factors that Affect Manganese Uptake, Distribution, and Utilization

Manganese uptake may be decreased by high pH or by high concentrations of calcium, magnesium, ammonium ion, iron, or other competing cations in the growth medium.[518] Molybdenum also affects the uptake of both manganese and iron and, hence, their interactions.[283] In general, the manganese concentrations in plant tops are good indicators of the manganese status of crops.[293] Manganese content is generally higher in leaves than in stems and petioles[283,356,432] and higher in mature leaves than in younger leaves[313] (Table 3-1). Manganese tends to accumulate in leaf margins, distal interveinal areas, leaf tips, and localized spots in older leaves.[263,544] Roots generally contain more manganese than do leaves.[528 (p.87)] Loper and Smith[318] noted that manganese concentrations in alfalfa decrease with maturity; a similar finding has been reported for lupines.[317] Gorsline et al.[204] concluded that corn absorbs 50% of its total manganese before the silking stage. Fergus[164] observed that manganese toxicity was more severe in young plants.

Although manganese and iron metabolism appear closely related in plants,[365] the two elements behave very differently in stem exudates under electrophoresis. Iron binds largely to citrate and moves as an anion, but manganese behaves as a cation and does not bind to citrate.[489] According to stability-constant calculations,[489] high calcium or magnesium concentrations in the xylem fluid can prevent the formation of manganese citrate. Bremner and Knight[73] reported that manganese in rye grass extracts occurred as a single cationic, probably noncomplexed form. Williams[547] suggested that a coordination of manganese to oxygen in oxyanions is the most probable type of binding. Höfner[236] proposed that manganese and iron in sunflower are chelated by amino acids and carbohydrates having molecular weights below 1,500.

Manganese Deficiency

Manganese deficiency is most common in small grains and soybeans, but also occurs in many other crops, including corn, peanut, cotton, sweet potato, sugarbeet, potato, sorghum, and mint.[463] Physiologic disorders associated with the deficiency include "grey speck" of oats, "marsh spot" of peas, and "speckled yellows" of sugarbeets.[293] Interveinal chlorosis in young leaves is often an early symptom of the deficiency. For details on manganese deficiency symptoms in various plants, see Mulder and Gerretsen,[365] Labanauskas,[293] and Jackson.[256]

For a rather wide range of plants, manganese deficiency has been reported when plant tops contained less manganese than about 20 ppm on a dry-weight basis, and concentrations of 20-500 ppm have been considered adequate but not toxic.[293] However, any such "critical" values vary widely with climate and with plant species or variety. Table 3-3 shows manganese concentrations that have been associated with deficiency, adequacy, or toxicity in plants.

Manganese Toxicity

Manganese toxicity in plants is characterized by marginal chlorosis and cupping of young leaves and speckling of older leaves, which is associated wtih localized manganese accumulations.[256,293] Williams and Vlamis[544] found that such necrotic spots in barley leaves were prevented by adding soluble silicon to the nutrient solution. The beneficial effect of silicon was attributed to a redistribution of manganese within the plant, rather than to reduced manganese uptake. Other physiologic disorders associated with manganese toxicity are "crinkle leaf" of cotton,[6] "stem streak necrosis" of Irish potato,[50,388] and "internal bark necrosis" of apple trees.[166,448] In severe cases of manganese toxicity, plant roots turn brown; but this generally occurs only after the tops have been noticeably injured.

Internal manganese concentrations associated with toxicity in plant tops vary widely with climate and plant species or variety. Ouellette and Dessureaux[386] observed toxic symptoms in alfalfa tops containing manganese at more than 175 ppm (Table 3-4). Foy[176] noted that manganese concentrations of 140 ppm or greater were associated with yield reduction in alfalfa grown in sand culture (Table 3-5). Pima S-2 cotton accumulated manganese at 1,740 ppm in its tops before toxicity symptoms developed and 2,000-5,000 ppm before yields were reduced[178] (Table 3-6). Leaf-bronzing symptoms occurred when carrot tops contained 2,600 ppm, and yields were reduced at internal concentrations of 7,100-9,600 ppm[209] (Table 3-3). The concentrations of manganese in the tops of manganese-sensitive plants are generally well correlated with injury and soluble manganese content in the growth medium, if other environmental factors are relatively constant.

Manganese toxicity can be reduced by increasing the concentrations of other cations (calcium, zinc, copper, and magnesium) that compete for absorption by plants in the growth medium.[139,253] In addition, some elements appear to interact with manganese inside the plant and thereby affect its toxicity. Manganese toxicity has been associated with iron deficiency in pineapple.[228] The addition of iron salts to the growth medium can reduce manganese toxicity in tobacco, rice, and clover. Some, but not all, of the manganese toxicity symptoms in flax

TABLE 3-3 Deficient, Adequate, and Toxic Manganese Concentrations in Plants

Plant	Type of Culture	Tissue Sampled	Other Information	Manganese Concentration, ppm (dry wt)			Reference
				Deficient	Adequate or Nontoxic	Toxic	
HORTICULTURAL CROPS							
Almond	Field	Leaves	—	5–25	96	—	155
Apple	Field	Leaves	—	15	30	—	376
Apple	Soil	Leaves	Interveinal bark necrosis	—	—	>400	166
Apricot	Field	Leaves	—	10	86–94	—	63
Avocado	Solution	Leaves	—	—	1,300	4,300–6,000	528
Avocado	Field	Leaves	September	—	366–655	—	294
Banana	Field	Leaves	—	<10	—	—	267
Bean (*Phaseolus vulgaris* L.)	Podzol soil	Tops	—	—	—	1,000	537
Bean (*Phaseolus vulgaris* L.)	Podzol soil	Leaves	pH, 4.7	—	—	600–800	255
Bean (*Phaseolus vulgaris* L.)	Field soil	Tops	ph, 4.7	—	40–940	1,104–4,201	316
Brussels sprouts	Field	Leaves plus petioles	Youngest fully expanded	—	78–148	760–2,035	308
Carrot	Solution	Tops	Bronzed leaves	—	—	>2,600	209
Carrot	Solution	Tops	Reduced yields	—	—	7,100–9,600	209
Lettuce	Steamed soil	Tops	Necrotic spots on leaves	—	—	>200	540
Lima beans	Field	Tops	—	32–68	207–1,340	—	392
Onion	Limed organic soil	Tops	Maturity	—	34	—	287
Orange	Field	Leaves	4–7 months	15	25–200	1,000	417

Orange	Field	Leaves	7-month bloom cycle, leaves from nonfruiting terminals	<15	20-50	>100	
Peas (*Pisum sativum*)	Podzol soil	Leaves	pH, 4.7	—	—	550	537
Pecan	Field	Leaflets	Detergent-washed	—	141–196	—	458
Potato	Field	Leaves		7	40	—	376
Potato	Soil	Leaves		—	—	473–2,290	103
Spinach	Field	Plant		23	34–60	—	196
Strawberry	Sand	Plant		—	50	—	254
Tomato	Solution	Leaves		5–6	70–398	—	321
Tung	Field	Leaves		58–81	—	—	142
Turnip	Field	Leaves		—	75	—	376
CEREALS							
Barley	Solution	Old leaves	Brown spots	—	—	1,200	520
Barley	Solution	Old leaves	Moderate to severe necrosis	—	—	305–410	543
Barley	Podzol soil	Leaves		—	—	200	537
Barley	Acid soil	Tops	pH, 4.7 Toxicity symptoms; yield reduced	—	—	80–100 700–1,000	415 415
Barley	Soil	Tops		—	14–76	—	380
Oats	Soil	Tops		8–12	30–43	—	380
Oats	Acid soil	Tops		—	301–370	—	316
Oats	Solution	Tops		—	774–783	—	315
Rice	Solution	Tops		<20	—	>2,500	478
Rice	Solution	Single leaf		—	—	4,000–8,000	478
Rice	Solution	Old leaves		—	—	7,000	520
Rye	Soil	Mature tops		—	10–50	—	265
Rye	Solution	Old leaves		—	—	1,400	520
Wheat	Soil	Tops	var. Thatcher	—	14–65	—	380
Wheat	Soil	Tops	var. Saunders	14	101	—	380
Wheat	Field	Plants		4–10	75	—	186
Wheat	Soil (pots)	Tops	Plants grown at soil pH 4.9 and 6.9	—	108–113	356–432	179

TABLE 3-3 (Continued)

Plant	Type of Culture	Tissue Sampled	Other Information	Manganese Concentration, ppm (dry wt)			Reference
				Deficient	Adequate or Nontoxic	Toxic	
Wheat	Solution	Tops	Plants grown with different manganese contents	—	181–621	396–2,561	179
Wheat	Solution	Roots	Plants grown with different manganese contents	—	1,094–6,183	4,096–11,761	179
Wheat	Solution	Tops	—	—	378–493	—	315
Corn	Field	Ear leaf	Single cross hybrids	—	116–214	—	204
Corn	Soil	Whole leaf	Moderately fertile soil	—	76–213	—	263
Corn	Field	Ear leaf	Several hybrids	—	19–84	—	37
Cotton	Sand	Leaf blade	105 days	<15	75–90	—	195
Cotton	Solution	Leaves	135 days	<10	10–1,200	—	482
Cotton	Soils	Tops	var. Pima S-2	—	27–216	1,130–2,920	178
Cotton	Solution	Tops	var. Pima S-2	—	196–924	1,740–8,570	178
Cotton	Solution	Tops	var. Rex Smooth Leaf	—	243–3,700	6,480	178
Cotton	Solution	Tops	var. Coker 100A	—	189	3,210	178
Cotton	Solution	Tops	var. Acala 4-42	—	216	2,810	178
Cotton	Solution	Tops	—	14	—	2,000	260
Cotton	Field	Leaves and petioles	16 varieties on 3 soils	—	58–238	—	18
Cowpeas	Solution	Tops	Toxicity symptoms	—	—	1,224	362
Cowpeas	Solution	Tops	Reduced yields	—	—	4,212	362
Hops	Solution	Leaves	—	—	37–88	903–1,796	533
Peanut	Solution	Tops	Slight toxicity symptoms	—	—	1,245	362

Crop	Medium	Plant part	Notes				
Peanut	Sand	Leaves	—	—	110–440	890–10,900	305
Soybean	Soil	Uppermost fully developed leaf	—	<20	—	—	119
Soybean	Soil	Tops	—	<15	35	—	375
Soybean	Solution	Leaves	—	2–3	14–102	173–199	461
Soybean	Solution	Tops	30 days; var. Earlyana	—	—	529	362
Soybean	Solution	Tops	Toxicity symptoms	—	—	2,168	362
Sugarbeet	Field	Leaves	Reduced yields	5–30	7–1,700	1,250–3,020	211
Sugarbeet	Field	Leaf blade	—	4–20	—	—	440
Sugarbeet	Field	3rd, 4th, 5th, and 6th blades from top of plant	—	1–10	20–400	—	440
Tobacco	Soil	Upper leaf	var. Burley 21	10	45	—	313
Tobacco	Solution	Tops	Burley	—	—	3,000	230
Tobacco	Solution	Top leaves	—	—	—	5,250–10,670	257
FORAGE CROPS—TEMPERATE GROUP							
Alfalfa	Solution	Tops	—	<10	—	—	46
Alfalfa	Field	Tops	—	—	—	477–1,083	316
Alfalfa	Solution	Tops	—	—	—	380	19
Alfalfa	Soil (pots)	Tops	Several soils of southeastern United States	—	65–240	651–1,970	177
Alfalfa	Sand	Tops	—	—	—	175–400	386
Alfalfa	Sand	Tops	—	—	110	140–2,200	176
Alfalfa	Solution	Tops	—	—	71–78	184–466	134
Barrel medic	Soil	Top	Cultivar 173	—	—	560[a]	19
Lespedeza	Soil	Top	—	—	—	>570	362
Ryegrass	Solution	Old leaves	—	—	—	800	520
Sweet clover	Solution	Tops	Toxic symptoms	—	—	321–754	362
Vetch	Field	Tops	—	—	—	500–1,117	316
White clover	Soil	Tops	var. New Zealand certified	—	—	650	19

TABLE 3-3 (*Continued*)

Plant	Type of Culture	Tissue Sampled	Other Information	Manganese Concentration, ppm (dry wt)			Reference
				Deficient	Adequate or Nontoxic	Toxic	
Medicago truncatula	Solution	Tops	Cultivar 173	—	—	560[a]	19
Trifolium fragiferum	Solution	Tops	Cultivar Palestine	—	—	510[a]	19
FORAGE CROPS—TROPICAL GROUP							
Trifolium subterraneum	Soil (pots)	Tops	—	4–25	30–300	—	498
Trifolium subterraneum	Solution	Tops	—	—	200	—	425
Centrosema pubescens	Solution	Tops	pH, 5.5	—	—	1,600[a]	19
Lotononis bainesii	Solution	Tops	Cultivar Miles	—	—	1,320[a]	19
Desmodium uncinatum	Solution	Tops	Cultivar Silver Leaf	—	—	1,160[a]	19
Stylosanthes humilis	Solution	Tops	—	—	—	1,140[a]	19
Phaseolus lathyroides	Solution	Tops	Cultivar Murray	—	—	840[a]	19
Phaseolus atropurpureus	Solution	Tops	—	—	—	810[a]	19
Glycine javanica	Solution	Tops	Cultivar Jineroo	—	—	560[a]	19
Leucaena leucocephala	Solution	Tops	—	—	—	550[a]	19

[a] "Toxicity threshold levels," defined as manganese concentrations found when yields were 5% below the maximum.

TABLE 3-4 Manganese Toxicity and Manganese Concentration in Alfalfa Tops Grown in Sand Culture[a]

Degree of Manganese Toxicity	Manganese Concentration in Stems and Leaves, ppm (dry wt)
None	<175
Very light (chlorosis)	175–250
Light (chlorosis and cupping)	250–325
Medium (crinkling of some leaves)	325–400
Severe (crinkling of most leaves)	>400

[a] Derived from Ouellette and Dessureaux.[386]

TABLE 3-5 Manganese Toxicity and Manganese and Iron Concentrations in Alfalfa Tops Grown in Sand Cultures[a]

Manganese Added in Solution, ppm	Relative Yield of Tops	Manganese Concentration in Tops, ppm	Iron Concentration in Tops, ppm	Iron:Manganese Ratio in Tops
0	83	100	375	3.75:1
0.5	100	110	300	2.73:1
1.0	89	140	250	1.79:1
4.0	83	220	270	1.23:1
8.0	73	250	290	1.16:1
32.0	45	1,200	290	0.24:1
64.0	25	2,200	280	0.13:1

[a] Derived from Foy.[176]

TABLE 3-6 Manganese Toxicity of Pima S-2 Cotton and Manganese Concentration in Plant Tops (Solution Culture Experiment)[a]

Manganese Added, ppm	Plant Yield, g/pot		Manganese Concentration, ppm		Degree of Crinkle Leaf Symptoms
	Tops[b]	Roots	Tops[b]	Roots	
0	1.71	0.31	196	112	None
2	1.61	0.35	924	336	None
4	2.01	0.46	1,740	728	Very slight
8	1.95	0.46	2,020	1,010	Slight
16	0.78	0.25	5,150	2,580	Moderate
32	0.25	0.17	8,570	5,630	Severe
64	0.00	0.08	—	—	Growing point dead

[a] Derived from Foy et al.[178]
[b] Tops above cotyledons.

are corrected by iron treatments.[365] Morris and Pierre[363] found that iron treatments reduce manganese toxicity in nutrient solutions but attributed the benefit to a reduction in manganese uptake, rather than an increase in iron uptake. They also reported that manganese toxicity symptoms in soybean, cowpea, and lespedeza appeared to be entirely different from those of iron deficiency.

Some investigators have emphasized iron:manganese ratios in relation to manganese toxicity and iron deficiency. In general, ratios between 1.5:1 and 2.5:1 have been considered optimal for plant growth.[365] For example, Rees and Sidrak[414] found that iron:manganese ratios of 3.08:1, 1.25:1, and 0.25:1 to 0.11:1 in barley were associated with no manganese toxicity, mild toxicity, and severe toxicity, respectively. In sand culture studies with alfalfa, Foy[176] found that yields were maximal when the ratio in plant tops was 2.73:1 and decreased steadily to 25% of the maximum as this ratio decreased to 0.13:1 (Table 3-5). Manganese toxicity symptoms (chlorosis and necrosis of leaf margins) appeared when plant tops contained manganese at over 250 ppm, but growth was reduced at 140 ppm. Manganese toxicity reduced the calcium concentrations of plant roots to half that of normal plants and greatly increased the manganese contents of these roots. However, manganese injury was not associated with changes in the concentrations of calcium, magnesium, phosphorus, boron, or iron in plant tops.

Rippel[420] concluded that excess manganese affected the action of iron within plants, rather than iron uptake. Hopkins et al.[238] concluded that the iron:manganese ratio controlled the growth of pineapple plants in high-manganese soils but found that, at a given ratio, the growth also varied with total concentrations of both iron and manganese in plant tissues. Misra and Mishra[357] attributed the beneficial effects of iron treatment to the oxidation of manganese within the plant. Watson[533] found that manganese-induced chlorosis in tomato was actually associated with increases in the total iron content of leaves and suggested that manganese interferes with iron metabolism. Singh et al.[453] found that both the height and the weight of soybean plants were influenced by the concentrations of iron and manganese but not by the ratios of their concentrations. Reid[415] reported that iron deficiency was not identical with manganese toxicity in barley and that the addition of iron to nutrient solutions or plant foliage did not reduce manganese toxicity symptoms or affect growth.

Millikan[355] found that molybdenum reduced manganese toxicity of flax in nutrient cultures and soils, but Gerloff et al.[193] reported that molybdenum at 0.67 ppm in the growth medium accentuated manganese-induced iron chlorosis, decreased iron uptake, and decreased growth in tomato.

Bortner[68] and Foy[177] obtained evidence that a high internal phosphorus concentration in plants may decrease the toxicity of a given internal manganese concentration. However, Morris and Pierre[363] observed no beneficial effects of phosphorus in relieving manganese toxicity.

Williams[545] noted that abundant manganese in the growth medium increased the severity of nickel toxicity in oats. The two metals apparently interacted to increase the rate of iron uptake but inhibited iron metabolism.

Two other factors reported to affect manganese uptake and toxicity are temperature and light intensity. Several investigators have found that higher temperatures increased manganese uptake by plants.[134,154,415,455,543] Epstein[154] suggested that the lack of severe manganese toxicity in Maine potatoes is due to reduced manganese uptake at low soil temperature. Munns et al.[366] showed that seasonal effects on manganese uptake coincided with temperature effects. There is also evidence that the toxicity of a given internal manganese concentration is affected by temperature and light intensity. For example, Löhnis[316] found that plants grown in a warm greenhouse tolerated higher manganese concentrations in their tops than those grown in the field; however, in this case, factors other than light and temperature may have been involved.

In apparent contrast with some of the results just cited, Sutton and Hallsworth[471] reported that increasing the growth-chamber temperature from 15 to 25 C had little effect on the tolerance of alfalfa to excess manganese in agar cultures. The manganese contents of alfalfa seedlings increased at low light intensities, but this had little effect on growth. High light intensities in combination with high manganese concentrations decreased growth and changed the toxicity symptoms from an orange-brown spotting to a chlorotic pattern resembling iron deficiency. Fujiwara and Ishida[185] found that the manganese uptake of rice plants grown under low temperature and shaded conditions was twice that of control plants.

Manganese toxicity in cotton has been associated with an increase in indoleacetic acid (auxin) oxidase activity, which results in the destruction of auxin.[360] Stonier et al.[467] reported that excess manganese destroys auxin protectors in Japanese morning glory.

Robinson and Hodgson[423] found that the addition of some amino acids to nutrient cultures decreased manganese toxicity in Irish potato. Methionine, cysteine, cystine, lysine, and tyrosine were effective, but alanine, valine, asparagine, and glutamine were not. The reduction of manganese toxicity by cystine was not associated with reduced manganese uptake. The data suggested that excess manganese created an imbalance in the amino acid pool of plants.

Anderson and Evans[14] found that manganese toxicity increased the activities of isocitric dehydrogenase and malic enzyme in leaves of bean. The addition of aluminum and iron depressed the activities of these enzymes in plant extracts. The malic enzyme activity of roots was decreased by toxic manganese concentrations and increased by iron and aluminum.

MANGANESE REQUIREMENTS

Economic plant species differ widely in manganese uptake under conditions of supposed optimal manganese availability in nutrient solutions[100] (Table 3-7). Under these conditions, peas, lettuce, and sunflower accumulated manganese concentrations that were five to six times those of tomato. Table 3-8 shows that orchardgrass accumulates much higher manganese concentrations than alfalfa, lespedeza, and red clover when grown on two Virginia soils.[409] Gerloff et al.[192] found that native plant species of Wisconsin differed even more widely in manganese concentrations (116–2,999 ppm) when all were grown in a bog soil at a pH of 4.0 (Table 3-9). Wallace[528 (p.18)] found that the iron-efficient Hawkeye soybean variety accumulated higher manganese concentrations in its stems and leaves than did the iron-inefficient PI 54619-5-1 variety. Bush beans were more efficient than corn in extracting easily reducible manganese

TABLE 3-7 Manganese Concentrations in the Tops of Different Plants Grown in Nutrient Solutions Containing Manganese at 0.55 ppm[a]

Plant	Manganese Concentration in Tops, ppm
Tomato	242
Vetch	384
Oats	741
Buckwheat	878
Spinach	922
Peas	1,367
Lettuce	1,378
Sunflower	1,521

[a] Derived from Collander[100] and Jackson.[256]

TABLE 3-8 Manganese Concentrations of Forage Plants Grown on Two Soils of Virginia[a]

Plant	Soil belt	Manganese Concentration in Tops, ppm (dry wt)	
		Range	Average
Alfalfa	Penn-Bucks	25-113	70.0
	Tatum-Nason	22-35	26.8
Lespedeza	Penn-Bucks	45-154	86.1
	Tatum-Nason	27-120	67.1
Red clover	Penn-Bucks	59-85	69.0
	Tatum-Nason	59-75	63.0
Orchardgrass	Penn-Bucks	298-368	333.0
	Tatum-Nason	73-168	122.0

[a] Derived from Price et al.[409]

from three soils.[529] Single-cross corn varieties have been reported to differ in manganese-accumulating ability.[204] Such differences may or may not reflect actual differences in manganese requirement. Lockman[313] reported differential manganese uptake by three tobacco varieties (Table 3-2).

Plant species and varieties also differ in the ability to absorb manga-

TABLE 3-9 Manganese Concentrations in the Tops of Native Wisconsin Plants Collected from a Bog Soil at a pH of 4.0[a]

Plant	Common Name	Manganese Concentration, ppm
Gaultheria hispidula	Creeping snowberry	2,999
Vaccinium myrtilloides	Velvet-leafed blueberry	2,177
Vaccinium oxyococcus	Small cranberry	1,340
Smilacina trifolia	False Solomon's seal	1,288
Chamaedaphne calyculata	Leather leaf	772
Dryopteris cristata	Crested-shield fern	426
Cornus canadensis	Bunchberry	149
Kalmia polifolia	Pale laurel	116

[a] Derived from Gerloff et al.[192]

nese from growth media that contain low concentrations. Loneragan et al.[317] found that the lupines accumulated much higher manganese concentrations in their tops than other species when grown on a lateritic, gravelly, low-manganese soil of Australia (Table 3-10). *Lupinus albus* (white lupine) was exceptionally efficient in accumulating manganese. Herbs and grasses had moderate to high manganese concentrations, and the cereals were moderate to low in manganese content. Species differences in manganese content were also found in straw, husks, and seeds (Table 3-10). In general, the abilities of plant species to accumulate manganese in their tops coincided with their abilities to grow on sandy low-fertility soils in which manganese deficiency is common. For example, rye and oats extracted manganese more effectively and had better growth on low-manganese soils than barley or wheat. Despite their differences in average manganese content, all the species showed similar relative changes in manganese concentration and total uptake when manganese fertilizers were added. The manganese contents of plant species were not related to root cation-exchange capacity or depth of rooting. Acid root excretions were believed to be important in regulating manganese uptake. The lupines showed a much greater net loss of manganese from their tops after flowering than did other species. This was attributed to translocation from the tops or loss from the foliage by rainfall leaching.

Anderson and Harrison[18] found that cotton varieties differed significantly in ability to extract manganese and iron from a Georgia soil with a pH of 6.0. Manganese and iron uptakes by 16 varieties were positively and significantly correlated.

Nyborg[380] demonstrated that cereal species and varieties differed in susceptibility to manganese deficiency in a Canadian soil that contained exchangeable manganese at less than 1 ppm (extracted with 1 N ammonium acetate at a pH of 7.0). Susceptibility ratings based on plant symptoms and yields were as follows: Olli and Parkland barley<Thatcher and Saunders wheat<Glen oats<Exeter, Victory, and Abeqweit oats. With no manganese added to the soil, the yields of species and varieties ranged from 37% to 100% of the maximum obtained with manganese fertilization. Manganese concentrations in plant tops ranged from 8 to 19 ppm, with the order as follows: oats<wheat<barley. Lack of manganese deficiency symptoms in barley was attributed to more effective manganese uptake. Greater sensitivity of oats to a low soil manganese concentration was ascribed to a reduced ability to absorb manganese, rather than a higher requirement for manganese. This ranking of oats and barley with respect to manganese feeding power in a Canadian soil is the opposite of that found by Loneragan et al.[317] for Australian soils.

TABLE 3-10 Manganese Concentrations in Legumes, Herbs, Cereals, and Grasses Grown on a Lateritic, Gravelly Sand at a pH of 5.0 in Australia[a,b]

Species or Cultivar	Manganese Concentration, ppm (dry wt)[c]			
	Plant tops[d]	Straw	Husks	Seeds
LEGUMES				
Lupinus albus	482	149	139	167
Lupinus digitatus	282	88	67	62
Lupinus angustifolius	246	30	52	29
Lupinus luteus	165	43	29	44
Ornithopus sativus	104	–	–	–
Ornithopus compressa	73	–	–	–
Vicia atropurpurea	55	–	15	29
Pisum arvense	50	–	21	–
Trifolium subterraneum (var. Bacchus Marsh)	43	–	–	32
Trifolium subterraneum (var. Yarloop)	42	–	–	23
Trifolium subterraneum (var. Clare)	36	–	–	25
Trifolium hirtum	37	–	–	16
Medicago truncatula	33	–	–	17
HERBS				
Cryptostemma calendula	101	–	–	–
Erodium botrys	77	–	–	–
CEREALS				
Secale cereale	113	54	–	43
Avena sativa (var. Ballidu)	73	40	–	39
Avena sativa (var. Avon)	57	33	–	34
Hordeum sativum	35	22	–	15
Triticum aestivum	35	35	–	34
GRASSES[e]				
Lolium rigidum	187(81)	–	–	–
Bromus rigidus	139(47)	–	–	–
Vulpia sp.	88[f](54)	–	–	–
Bromus mollis	67[f](33)	–	–	–

[a] Derived from Loneragan et al.[317]
[b] pH measured in a 1:5 soil–0.01 M calcium chloride suspension.
[c] Average of three harvests and three micronutrient treatments that included 6.4 and 19.1 kg of $MnSO_4 \cdot 4H_2O$ per acre.
[d] From preflowering to postflowering stages.
[e] Numbers in parentheses indicate concentrations in whole tops of mature plants.
[f] Believed to be borderline manganese-deficiency content; however, yield was not reduced.

The difference could be due to differences in soil, climate, or crop variety.

Differences in tolerance to low manganese concentration in the growth medium have been reported previously for varieties of oats,[366, 522,523] wheat,[373] and rice.[312] Gallagher and Walsh[186] suggested that manganese deficiency could be largely overcome in cereals by selecting varieties that are more efficient in extracting manganese from the soil. Munns et al.[366] reported that differences in manganese uptake among oat variety tops persisted under different seasonal conditions and under variable substrate conditions, including pH, calcium concentration, iron supply, nitrogen source, and manganese concentration. Such differences in manganese uptake, which persisted even in mixed cultures, were attributed to internal factors, rather than to the effects of plants on the nutrient solution. Varietal differences in shoot contents of manganese were attributed to variations in the size and turnover rates of a labile manganese fraction in the roots. Munns et al.[367] also found that temperature affected manganese uptake and distribution in oats but did not alter varietal differences in manganese contents of tops. Higher concentrations of manganese in the roots of one variety at low temperature or high pH were attributed to accumulation of manganese in a nonlabile form. In this connection, Kleese and Smith[284] found that genetic differences in manganese accumulation by soybean seeds are controlled by manganese movement in the shoot, rather than in the root.

Hasler,[215] cited by Mulder and Gerretsen,[365] classified 15 grass species according to their response to manganese fertilization. Species showing the greatest response were *Alopecurus aratensis, Arrhenatherum elatius,* and *Bromus erectus;* those showing intermediate responses were *Anthoxanthum odoratum, Cynosurus critatus, Festuca pratensis,* and *Trisetum flavescens;* and those showing little or no yield response to added manganese were *Agrostis alba, Dactylis glomerata, Festuca rubra, Lolium italicum, Lolium perenne,* and *Poa pratensis.* Root systems of manganese-deficient grasses were poorly developed.

TOLERANCE TO EXCESS MANGANESE

Range of Tolerance

Plant species and varieties differ widely in their tolerance to excess manganese in soils and nutrient solutions.[229, 256] Table 3-11 shows that oats and alsike clover are more tolerant than kale, tomato, or alfalfa. Peanuts are more tolerant than lespedeza and sweet clover.[362] Kobe les-

TABLE 3-11 Differential Tolerance of Plant Species to Excess Manganese[a]

Plant	Relative Yield, %[b]
Kale	30
Tomato	40
Alfalfa	40
Brussels sprouts	45
Carrot	50
Red clover	60
Cauliflower	60
Celery	65
Barley	70
Leek	70
Sugarbeet	85
Potato tubers	90
Alsike clover	100
Oats	100

[a] Derived from Hewitt[229] and Jackson.[256]
[b] Yield after treatment with excess manganese as a percentage of yield after optimal manganese treatment.

pedeza, black locust, and birdsfoot trefoil resist manganese toxicity better than Korean, sericea, or bicolor lespedeza.[49] Löhnis[316] showed that *Trifolium pratense, Trifolium repens,* and *Vicia faba* were more tolerant to excess manganese than *Phaseolus vulgaris, Vicia sativa,* and *Medicago sativa.* Robson and Loneragan[425] found two cultivars of *Trifolium subterraneum* far more tolerant to excess manganese than two annual Medicago species. Differential manganese tolerance among species of tropical legumes has also been reported by Andrew and Hegarty[19] (Table 3-3). The ornamental plants calendula, snapdragon, and chrysanthemum are sensitive to high manganese concentration in the growth medium and have been suggested as indicator plants,[102] but carnation, poinsettia, and rose are classified as manganese-tolerant.

Ouellette[387] concluded that for normal growth the soil concentrations of soluble manganese (5 g of soil extracted with 100 ml of acetic acid, with a sodium acetate buffer) could not exceed 1.0, 1.5, 2.0, and 3.0 ppm for potatoes, clover, lespedeza, and soybeans, respectively. Manganese concentrations of 1–4 ppm in solution were harmful to lespedeza, soybeans, and barley,[362,383] but corn tolerated over 15 ppm, and *Deschampsia fluxuosa,* over 60 ppm.[383] Manganese toxicity has been observed in nutrient solutions containing 15 ppm for tobacco,[68] 10 ppm for cotton,[6] and less than 10 ppm for several legumes.[362] Alfalfa has been injured in sand cultures when the saturating nutrient culture contained 4–8 ppm,[176] and similar concentrations in nutrient solution produced manganese toxicity (crinkle leaf) in Pima S-2 cotton.[178]

Williams and Vlamis[543] found that the recommended manganese concentration in Hoagland's solution (0.5 ppm) was toxic to barley, slightly toxic to lettuce, and nontoxic to tomato. These results for barley and tomato are opposite those of Hewitt[229] (Table 3-11) but could be explained by differences in plant variety or other experimental conditions. In later work, Vlamis and Williams[520] showed the following order of manganese tolerance: ryegrass and rice>rye>barley. Differences in manganese tolerance have also been shown for varieties of cotton,[178] alfalfa,[386] Irish potatoes,[388] and wheat.[373]

Physiologic Characterization of Differential Tolerance

EXCLUSION FROM PLANT TOPS

Some manganese-tolerant plants appear to escape injury either by absorbing less manganese or by trapping the excess in the roots or other plant parts, where it is physically or chemically separated from key metabolic sites. Resistance to manganese toxicity in some rice varieties has been attributed to lower manganese absorption.[312] Ouellette and Dessureaux[386] found that manganese-tolerant alfalfa clones contained lower concentrations of manganese in their tops and higher concentrations of manganese and calcium in thier roots than did the manganese-sensitive clones. These investigators concluded that calcium uptake regulated manganese toxicity by reducing the transport of manganese to plant tops. Manganese tolerance in some species of tropical legumes also seems to depend in part on the retention of excess manganese within the root system and the prevention of toxic accumulations in plant tops.[19] Robson and Loneragan[425] found that *Trifolium subterraneum* was more tolerant to excess manganese than some Medicago species of annual tropical legumes. The greater tolerance of the former was associated with a lower rate of manganese absorption per gram of root weight and a greater retention of manganese by the roots. Lunt and Kofranek[320] suggested that manganese may be detoxified by accumulating in the woody portion of the azalea plant. Ouellette and Genereaux[388] noted that the leaves of manganese-tolerant Irish potato varieties accumulated lower concentrations of manganese than did those of manganese-sensitive varieties. Gerloff et al.[192] reported that some native plant species of Wisconsin selectively exclude manganese, at least from their tops (Table 3-9). Heintz[222] concluded that manganese was detoxified by precipita-

Manganese and Plants

tion with phosphorus in plant roots. Sutcliffe[469] suggested that roots contain organic complexes that render manganese unsuitable for transport to plant tops.

Waterlogging of soils promotes the reduction of manganese to the divalent (available) form, and there is some evidence that the tolerance of plant species to wet soils coincides with tolerance to excess manganese. Graven et al.[206] suggested that the wet-soil sensitivity of alfalfa might be due in part to manganese sensitivity. They also pointed out that pasture species most tolerant of waterlogging[167,344] are generally also tolerant of high manganese concentrations in nutrient solutions.[229,362] Robson and Loneragan[425] noted that manganese-tolerant *Trifolium subterraneum* (var. Geraldton) was more tolerant to waterlogging than the manganese-sensitive *Medicago truncatula* and *Medicago littoralis* species.

Doi[137,138] suggested that tolerance of plant species to wet soils is associated with the oxidizing powers of their roots. Oxidizing values were highest in the Gramineae and Compositae, somewhat lower in the Leguminoseae, and lowest in the Cruciferea, Solanaceae, and vegetable crops. Soybean was injured in paddy soil when planted alone, but this injury was reduced when soybean was planted with rice. The higher oxidizing powers of the rice roots were believed to benefit the respiration of the soybean roots.

TOLERANCE TO HIGH INTERNAL CONCENTRATIONS

Some plants appear to tolerate high concentrations of manganese within their tops. Löhnis[315] concluded that some plants that are resistant to manganese toxicity take up large quantities of the element. For example, *Lupinus luteus* and *Ornithopus sativus*, both highly tolerant to excess manganese and acid soils, accumulate high concentrations of manganese in their tops. Löhnis[315] observed that *Vaccinium myrtillus* plants were healthy when the foliage contained manganese at 2,432 ppm. Vlamis and Williams[519] found that rice is much more tolerant to excess manganese than barley but still accumulates five to six times as much manganese in its leaves. Iron concentrations in rice leaves were double those in barley leaves. The cotton variety Rex Smooth Leaf, developed in the high-manganese soils of Arkansas, is more resistant to manganese toxicity than Coker 100A, which was developed in the low-manganese, sandy Coastal Plain soils of South Carolina.[178] Greater manganese tolerance of Rex Smooth Leaf has been attributed to an ability to tolerate higher manganese concentrations within its tops, rather than to reduced manganese uptake.

TOXIC CONCENTRATIONS IN PLANT TOPS

For a given plant, the severity of manganese toxicity is generally directly proportional to the manganese concentration in plant tops, but the internal concentration required for toxicity depends on many factors, particularly plant genotype. Table 3-3 shows a wide range of internal manganese concentrations that have been associated with toxicity in different plants. For example White[537] found that toxicity symptoms in beans, peas, and barley were associated with manganese concentrations of 1,000, 550, and 200 ppm, respectively. Toxicity in rice has been associated with manganese concentrations above 2,500 ppm in whole plant tops and as much as 4,000 or even 8,000 ppm in single leaves.[478] Buffalo alfalfa grown in a greenhouse in several acid and limed soils of the southeastern United States had good growth when the tops contained between 65 and 240 ppm, with most values falling between 100 and 200 ppm. Manganese toxicity was present at manganese concentrations between 320 and 1,970 ppm in plant tops.[177] In sand cultures, manganese toxicity has been associated with concentrations above the range of 140–175 ppm in alfalfa tops.[176,386] Because such so-called "critical" values vary with many factors, they must be determined for a well-defined set of conditions to be most useful.

REGULATING MANGANESE CONTENT

Soil Management

For a given crop variety, manganese deficiencies associated with low availability of soil manganese can often be prevented by keeping the soil pH below 6.5, particularly in organic, sandy soils. This can be accomplished by avoiding overliming or by adding elemental sulfur to reduce the soil pH to an acceptable point for optimal manganese solubility. For soils low in total or available manganese, banding 14–27 kg of manganese sulfate per acre is recommended at planting time.[293] Manganese deficiency in a growing crop can be corrected by foliar sprays of manganese sulfate.[293] According to Wallace,[528 (pp.46,215)] soil-applied manganese chelates are usually not effective sources of manganese for plants, particularly in acid, organic soils that are high in iron or copper.

Manganese toxicities in a given crop variety (or high manganese uptake without toxicity) can be prevented by liming the soil to a pH of 5.5–6.0, avoiding heavy fertilization of acid soils, adding organic matter, improving soil drainage, and avoiding soil compaction.

Selection of Plant Species or Variety

In soils low in available manganese, plant species or varieties efficient in extracting and using soil manganese should be used. Varietal selection is particularly important in such a species as oats, which is very susceptible to manganese deficiency.

In soils with a current or potential manganese-toxicity problem (because of high total manganese concentration, poor drainage, or the necessity for maintaining a low pH for some crops like potatoes), crop species and varieties can be used (or varieties developed) that either selectively exclude excess manganese from their tops or tolerate high internal manganese concentrations. For example, alfalfa is generally sensitive to manganese toxicity, but different varieties and clonal selections differ in manganese tolerance, and this tolerance is apparently genetically controlled.[135] The evidence indicates that the genes that regulate manganese tolerance are additive and have little or no dominance.[133]

A plant-breeding approach to the problem of manganese regulation in plants could be effective in maintaining crop yields and in ensuring that plant materials consumed by animals and man contain adequate but not excessive manganese for their diets. Plants that exclude excess manganese from their tops would be useful in avoiding excesses in crops grown in high-manganese environments. Plants that accumulate extremely high manganese concentrations in their tops without injury may be useful in detoxifying soil contaminated by excess manganese fertilizers or wastes.

SUMMARY

The availability of manganese to a given crop is affected by numerous soil factors acting independently or in conjunction with each other. These include concentrations of total or easily reducible manganese; pH; concentrations of other cations, anions, and total salts; cation-exchange capacity; organic-matter content; drainage; compaction; temperature; and microbial activity. Small changes or inherent differences in such factors can determine whether the soil manganese concentration is deficient, adequate, or toxic for a given crop. There are various soil tests for predicting manganese deficiency and toxicity.

The utilization of absorbed manganese by a given plant is also affected by many factors, such as temperature, light intensity, and interactions between manganese and iron or other elements.

Plant species and varieties differ widely in manganese requirement and accumulating ability and also in tolerance to excess manganese concentrations within their tissues.

Manganese contents and yields of crops can be regulated by soil-management practices that alter the concentration of divalent manganese in soils, by the application of manganese-containing foliar sprays, and by choice of crop species or variety. Evidence that manganese uptake and use are genetically controlled in some crops suggests the possibility of breeding varieties for greater ability to extract manganese at low concentrations, to exclude excess manganese from their tops, or to tolerate high manganese concentrations within their tissues. Such varieties would be useful in maintaining crop yields and in regulating manganese contents of plant products.

4

Input and Disposition of Manganese in Man

SOURCES AND ROUTES OF INPUT

Diet

Manganese is classed as an essential mineral[48,379,430,477] for man and animals. It has been shown to be necessary for normal bone formation and synthesis of chondroitin sulfate.[302] Deficiency states have been demonstrated in many species of animals, but unequivocal evidence of manganese deficiency, absolute or conditioned, has not been obtained in man; the National Academy of Sciences *Recommended Dietary Allowances*[371] does not list a recommendation for man. However, minimal dietary requirements have been established for most animal and avian species and have been observed to depend on species, criteria of adequacy, chemical form ingested, and nature of the remainder of the diet.[29,48,208,218,237,250,261,302,406,407,451,459,526]

The common foods consumed by man vary considerably in manganese content. Table 4-1 lists several major food groups in descending order of average manganese concentration on a fresh-tissue basis. Ranked on a dry-weight basis, the leafy vegetables would be much higher. Other investigators have confirmed and extended these earlier findings. From such analyses, it is apparent that the daily intake of manganese in hu-

TABLE 4-1 Manganese Content of Groups of Principal Foodstuffs[a]

Class of Food	No. Samples	Manganese Concentration, ppm (fresh-tissue basis)		
		Minimum	Maximum	Average
Nuts	10	6.3	41.7	22.7
Cereal products	23	0.5	91.1	20.2
Dried legume seeds	4	10.7	27.7	20.0
Green leafy vegetables	18	0.8	12.6	4.5
Dried fruits	7	1.5	6.7	3.3
Roots, tubers, and stalks	12	0.4	9.2	2.1
Fresh fruits (including blueberries)	26	0.2	44.4	3.7
Fresh fruits (excluding blueberries)	25	0.2	10.7	2.0
Nonleafy vegetables	5	0.8	2.4	1.5
Animal tissues	13	0.08	3.8	1.0
Poultry and byproducts	6	0.30	1.1	0.5
Dairy products	7	0.03	1.6	0.5
Fish and seafoods (excluding oysters)	7	0.12	2.2	0.5
Fish and seafoods (including oysters)	6	0.12	0.4	0.25

[a] Derived from Peterson and Skinner.[403]

man diets varies considerably with the nature of the diet, but the daily consumption on the average ranges from 3 to 7 mg. This amount is probably sufficient to prevent deficiency symptoms, but diets high in milk, sugar, and refined cereals might provide insufficient manganese, particularly in growing children, pregnant women, and persons with diabetes or rheumatoid arthritis,[113] in which a slow turnover rate of manganese has been demonstrated. Furthermore, these data indicate that dietary intake is unlikely to contribute to an excess body burden of manganese in man, although entry via the gastrointestinal tract can result in intoxication in miners who work with manganese ores.[104–106,140,221] Intoxication via drinking water has also been reported.[275]

Of the trace elements, manganese is among the least toxic to mammals and birds. Hens tolerate 1,000 ppm without ill effects,[187] but 4,800 ppm is toxic to young chicks.[224] Rats tolerate high concentrations of dietary manganese,[187] with growth rate unaffected by concentrations as high as 2,000 ppm. Swine tolerate somewhat lower concentrations; 500 ppm depresses appetite and retards growth.[208] Calves had decreased feed intake and lower body-weight gains when their diet was supplemented with manganese at either 2,460 or 4,920 ppm; supplements of 820 ppm had no detectable effect on growth or appetite.[120] Lambs,[214] cattle,[424] rabbits, and pigs[332] develop problems in maintaining hemoglobin concentrations when dietary manganese remains high—a phenomenon that appears to be related to iron metabolism.

Respiration

Inhalation appears to be the major route of absorption in cases of intoxication in man, although absorption through the skin is reported.[25,26,221] Intoxication via the respiratory route can result in chronic manganism, primarily a disease of the central nervous system often first manifested by disordered mentation. It results from exposure to high concentrations of manganese dust, often for only a few months. Such exposure can also result in manganic pneumonia, a form of lobar pneumonia unresponsive to treatment with antibiotics.[1,38,132,433,531,538] Although the average dose of manganese that causes pneumonia is not known, miners who worked with pneumatic drills in Chilean manganese mines in an atmosphere of approximately 5,000 particles/m^3 developed manganism in an average of 178 days.[531] Reports from Russia[140] and from Norway[150,151] indicate a high rate of inhalation toxicity in human populations near centers of steel production and manganese alloy manufacture. The similarity of symptoms of chronic manganese poisoning and parkinsonism has been detailed by Cotzias *et al.*[112]

ABSORPTION AND EXCRETION

The early studies of Greenberg *et al.*[207] with radiomanganese indicated that only about 4% of an orally administered dose was absorbed by rats. What was absorbed appeared quickly in the bile and was excreted in the feces. Experiments in animals and man have shown that injected radiomanganese disappears rapidly from the bloodstream.[66,274] The disappearance has been resolved into three phases: the first and fastest, which is identical in rate with the clearance of other small ions and is thus probably a normal transcapillary movement; the entrance of manganese into the mitochondria; and the slowest, nuclear accumulation. Inasmuch as the kinetic patterns for blood clearance and liver uptake in animals are almost identical, the two manganese pools must rapidly enter equilibrium, and body manganese must be in a highly mobile dynamic state. Actually, little is known about the mechanism of absorption of manganese from the gastrointestinal tract.

Most of the absorption data on manganese are based on animal studies;[221] the data on man are derived mainly from clinical observations and epidemiologic studies.[140,311,389,513]

Experiments in animals and man show conclusively that manganese is almost totally excreted via the gastrointestinal tract, being secreted through the intestinal wall and eliminated in the bile.[57,274,391,531] Under ordinary conditions, the bile flow is the main route of excretion and constitutes the principal regulatory mechanism. The routes appear to be interdependent; together, they constitute an efficient homeostatic mechanism that regulates manganese content in the tissues. The relative stability of tissue manganese concentrations is due to such controlled excretion, rather than to regulated absorption.[77]

Total manganese concentration in the bile is correlated with its bilirubin content[335] and can be increased by manganese loading. Excretion also occurs via pancreatic excretions[82] and via the duodenum, jejunum, and ileum,[57] the latter probably serving as auxiliary routes. Very little manganese is excreted in the urine.[277,335]

Mahoney and Small[325] have shown that the disappearance of labeled manganese from the human body can be described by a curve with two exponential components; 70% of injected manganese was eliminated via a pathway with an average half-time of 39 days, and the more rapid pathway had a half-time of 4 days.

ACCUMULATION AND DISTRIBUTION

It has been estimated that a normal 70-kg man has a total of 12–20 mg of manganese in his body.[105,106] Manganese is widely distributed throughout the body; concentrations are characteristic for the various organs and tissues and vary little within or among species.[443] Manganese is generally higher in tissues rich in mitochondria, and it tends to concentrate in the mitochondria.[334,486] Higher concentrations of manganese are generally associated with pigmented portions of the body,[114,481,515] including the retina, pigmented conjunctiva, dark hair, and dark skin.

Table 4-2 lists average concentrations of manganese in tissues of men, rabbits, and a variety of other animal species.[175,443,552] The pituitary gland, pancreas, liver, kidney, and bones normally have higher concentrations of manganese, and skeletal muscle has a very low concentration. Hair accumulates manganese in relatively high concentrations.[26] The storage capacity of the liver for manganese is limited and offers a contrast in this regard with iron and copper.[8,261,331] Furthermore, reserve stores of manganese are not present in the liver of the newborn of several species, including man.[500] Human livers from healthy people of all ages contain manganese at about 6–8 ppm (dry-weight basis).[79,447,491] In contrast with many other trace metals, manganese does not accumu-

TABLE 4-2 Concentrations of Manganese in Tissues of Man and Animals[a]

	Manganese Concentration, ppm (fresh-tissue basis)		
Tissue	Men	Rabbits	Average for Several Species
Adrenal	0.20	0.67	0.40
Aorta	0.19	–	–
Bones (long)	–	–	3.30
Brain	0.34	0.36	0.40
Heart	0.23	0.28	0.34
Kidney	0.93	1.20	1.20
Liver	1.68	2.10	2.50
Lung	0.34	–	–
Muscle	0.09	0.13	0.18
Ovary	0.19	0.60	0.55
Pancreas	1.21	1.60	1.90
Pituitary	–	2.40	2.50
Prostate	0.24	–	–
Spleen	0.22	0.22	0.40
Testis	0.19	0.36	0.50
Hair	–	0.99	0.80

[a] Derived from Underwood.[500]

late significantly in the lungs with age, averaging about 0.22 ppm in aged man.[359,443,488,491]

The only manganese-containing metalloprotein with a fixed concentration of manganese per molecule of protein is pyruvate carboxylase.[445] However, it is claimed that liver arginase contains manganese as an essential component.[36] In human serum, manganese is nearly totally bound to β_1-globulin.[56,110,174] In the liver, manganese is mainly in the arginase extract, which indicates that it is largely protein-bound.

SUMMARY

Manganese is classed as an essential mineral for man and animals. The usual diets consumed by man generally provide sufficient quantities to prevent a deficiency, although diets high in milk, sugar, and refined cereals may provide only minimal amounts of manganese for growing children, pregnant women, and diabetics. The dietary sources are therefore unlikely to contribute to an excess body burden in man.

Although manganese is among the trace elements least toxic to mammals, man-made pollution—particularly that related to steel industries—and exposure to the mining of manganese ore represent potential toxic hazards to persons working in uncontrolled areas.

Manganese can be absorbed by inhalation, by ingestion, and through the skin. Most damage results from inhalation, which can result in chronic manganese poisoning, a disease of the central nervous system, or in a form of pneumonia.

Little is known about absorption of manganese from the gastrointestinal tract, but excretion is normally via the biliary route in the feces. Secretion through the intestinal wall is also known to occur, with secretion apparently governing an efficient homeostatic mechanism, rather than regulated absorption.

Skeletal structures, hair, liver, pancreas, and kidney contain concentrations of manganese that are characteristic for the tissues but vary little among species. Other tissues are reported to contain lower concentrations. Within the cell, the highest concentrations of manganese are in the mitochondria. Generally, organs and tissues do not accumulate large concentrations of manganese.

Removal of a victim from the polluted environment in the early stages usually clears up manganism; in chronic cases, chelation removes the metal to some extent but appears to have little effect on central nervous system symptoms.

5

Biochemistry and Metabolic Role of Manganese

BIOCHEMISTRY

Manganese is of interest not only to the biochemist but to the physiologist, nutritionist, toxicologist, and physician, because results of its deficiency or toxicity are similar to some spontaneous human diseases whose causes have been difficult to determine.[546] Manganese occurs in minute concentrations within the cells of all living things and has been established as essential to a wide variety of organisms ranging from bacteria to plants and mammals.[548] However, chronic inhalation of manganese dusts can cause manganism in man, a syndrome characterized by progressive extrapyramidal manifestations, sometimes accompanied by an unrelated pneumonitis. Braunite appears much more poisonous than pyrolusite, and ores allowed to age lose toxicity along their surface, which suggests that the less oxidized is the compound the higher is its toxicity.[105]

The characteristic oxidation state of manganese in solution is Mn^{+2}, a d^5 ion.[381] Under the usual conditions, manganese complexes are composed of Mn^{+2} coordinated with six ligands to give an octahedral geometry.[381] The stability of many metal complexes in enzymatic reactions follows the ranking order known as the Irving–Williams series. Although Mn^{+2} forms diverse complexes, their formation constants are less than those of succeeding elements in the first transition series. Thus, one might predict the existence of fewer metalloenzymes that contain manganese than contain other transition elements.[381] In most synthetic complexes, manganese has five unpaired electrons and is consequently paramagnetic. Because of the resulting large magnetic moment, Mn^{+2} has a large effect on the longitudinal proton relaxation rate of water protons, the increased rate being effected by promotion of proton–electron dipolar magnetic interaction.[99,381]

An understanding of the interaction of metal ions with nucleotides and nucleic acids is greatly needed. The complex of adenosine triphosphate (ATP) and metal, because of its importance in a great many biologic reactions, has been of particular interest. Mn^{+2} and Mg^{+2} are two important metal ions associated with ATP. Mn^{+2} can replace Mg^{+2} in many biologic reactions of the metal-ATP complex. How or why this can occur needs to be understood. From water-proton relaxation times, progress has been made in showing the binding sites of Mn^{+2} in the nucleotide and the different structures for such complex metal-containing compounds. A detailed knowledge of the structure is a prerequisite to understanding the role of the metal in various enzymatic reactions.

The specific biochemical role of manganese has been difficult to determine. Manganous ion has long been recognized as an activator of enzymes that require the presence of a divalent ion, but generally the activation is nonspecific.[381] These nonspecific manganese-activated enzymes include hydrolases, kinases, decarboxylases, and transferases.[512] Several important enzymes demand manganese exclusively, including the peptidases prolidase and succinic dehydrogenase. Prolidase is an intestinal digestive enzyme that splits the dipeptide glycylproline, and succinic dehydrogenase is involved in the citric acid cycle.[385]

Clear evidence of a manganese metalloenzyme is provided by pyruvate carboxylase.[381,445] Pyruvate carboxylase is a mitrochondrial enzyme that catalyzes the carboxylation of oxaloacetate from pyruvate. Carboxylation reactions of this kind involve the participation of biotin, a coenzyme attached to the lysyl residues of the specific enzymes, which serves as a carrier for the carbon dioxide that is transferred.[340] Pyruvate carboxylase is unusual in requiring two metallic ions. Magnesium ion is required for the carboxylation of the biotinyllysyl residue, and manganous ion for binding the keto compounds to the active sites of the enzymes.[340] The fact that acetyl coenzyme A does not participate in the reaction in any way but is absolutely required as a modulating activator appears to be related to one of the key mechanisms of control in hepatic metabolism. Pyruvate can diffuse in and out of mitochondria, but acetyl coenzyme A cannot. The localization of the carboxylase step within the mitochondria and its sensitivity to acetyl coenzyme A make gluconeogenesis, which is a process mainly of the cytosol, sensitive to changes in oxidative metabolism.[340] Thus, the importance of manganese in carbohydrate metabolism is indicated by its presence in the metalloenzyme pyruvate carboxylase. To support this possibility, there is evidence implicating manganese in glucose utilization.[162,182,495]

The role of manganese in the function of other enzymes involved in carbon dioxide fixation also has been studied.[354,381] In addition, manganese has long been associated with arginase, although arginase cannot be classified as a metalloenzyme. Arginase activity in tissues is decreased by manganese deficiency, and arginase activity depends heavily on manganese.[234,381] The catalytic activity and stabilization of the quaternary structure of glutamine synthetase require Mn^{+2} or Mg^{+2}. The affinity of the enzyme for Mn^{+2} is 400 times that for Mg^{+2}.[131,381] Thus, it has not been possible to specify a single biochemical role for manganese. However, whether it has one or many biochemical roles, its function in many cases cannot be taken over by other metals.

FORMATION OF CONNECTIVE TISSUE AND BONE

Nutritional deficiency of manganese results in skeletal abnormalities in all species investigated. The type and severity of the deformity depend on age. Deficiency of this element in the embryo results in different, often more severe effects than deficiency in the growing animal after birth. For example, deficiency in the chick embryo results in nutritional chondrodystrophy—which is characterized by short, thick legs and wings, shortening of the lower mandible ("parrot beak"), and a skull that protrudes anteriorly—and in a high mortality rate; and deficiency in the chick causes perosis, characterized by lameness resulting from malformation of the tibiometatarsal joint, thickening and shortening of the long bones, and slipping of the gastrocnemius tendon from its condyles.

It is typical of manganese deficiency that some bones, especially of the appendicular skeleton,[27] are affected more than others; this results in disproportionality of skeletal growth.[90] In addition to being shortened, bones may be deformed and thickened; and thickened, malformed joints are common.

Skeletal deformities have also been reported in turkeys,[160] swans,[153] ducklings,[52] mice,[43] rabbits,[459] pigs,[353] sheep,[296] goats,[20] and cattle;[205] in each instance, the limb bones and joints were affected primarily. For example, manganese deficiency in growing swine causes stiffness, lameness, enlarged hocks, shortening and bowing of forelimbs, shortening of hind limbs, thickening of carpal and tarsal bones, and lipping of distal ends of the radius and ulna.[27] It has been reported that "overknuckling," a disease of cattle that is characterized by poor growth and bone deformity and occurs when cows graze in some pastures in Holland, is due to manganese deficiency,[205] although this has recently been questioned[51] (J. Hartmans, personal communication).

Congenital defects in the bones of offspring of manganese-deficient mothers have been observed in species other than chickens. In rats, maternal deficiency produces shortening of the radius, ulna, tibia, and fibula and shortening and doming of the skull. Hurley and Asling[245] have described a marked epiphyseal dysplasia of the proximal tibial epiphysis.

The observed skeletal disorders are associated with and secondary to impaired activity of epiphyseal cartilage plates. Histology of the cartilage[550] shows reduced chondrogenesis and faulty matrix formation. Although there is some decrease in density, breaking strength, and ash content of bone, the impairment of the calcification process is not a primary, causal factor in the bone abnormalities and, indeed, is not a major feature of the condition. The initial report of low blood and bone alkaline phosphatase concentrations in manganese-deficient animals[542] directed attention to the calcification process, but later reports have indicated that concentrations of this enzyme are not consistently reduced; moreover, the relation of such deficiency to these particular bone and cartilage abnormalities is not apparent.

When it became apparent that the primary changes were in the epiphyseal plate, the chemical composition of this plate was determined by Leach in 1960.[300] He found a substantial increase from normal in fat content and decrease in hexauronic acid content in manganese deficiency. Further investigations revealed a decrease in hexosamine,[301] a lowered sulfur-35 uptake, and a decreased percentage of galactosamine. These data indicated a severe reduction in cartilage chrondroitin sulfate content.[301] Histologic observations by Leach[299] revealed a loss of periodic acid-Schiff-positive (PAS-positive) extracellular matrix. Erway et al.[158] have reported a marked reduction in acid mucopolysaccharides in cartilage of manganese-deficient guinea pigs at birth.

Recently, Leach et al.[302] have defined the enzyme steps that are defective in manganese-deficient cartilage:

1. polymerization of uridine diphosphate-N-acetylgalactosamine to uridine diphosphate-N-acetylglucuronic acid, which is required for the synthesis of the polysaccharide; and

2. incorporation of galactose, by galactotransferase, from uridine diphosphate-N-acetylgalactose into the trisaccharide (galactose-galactose-xylose), which forms the linkage between the polysaccharide and a serine of the associated protein.

These findings not only provide a biochemical explanation for the observed defects in connective tissue and skeleton resulting from manganese deficiency, but constitute, in fact, the first established link between the observed effects of manganese deficiency *in vivo* and a specific underlying biochemical defect.

DEVELOPMENT OF THE INNER EAR

Ataxia—characterized by incoordination, loss of equilibrium, and head retraction—has been observed in the offspring of manganese-deficient chickens, rats, mice, and guinea pigs.[91,158,451] This cannot be reversed by manganese supplementation after birth, and there is a critical time late in gestation after which dietary maternal manganese supplementation is ineffective in preventing this lesion. In the rat, for example, maternal manganese supplementation of deficient animals before the fourteenth day of gestation prevents the neonatal ataxia, but supplementation on the eighteenth day is of no benefit.[248] Thus, the defect results from a maternal deficiency of manganese at a specific time relatively late in pregnancy. It was originally thought to result from a metabolic disorder in the brain, although no associated abnormalities of brain, spinal cord, or cerebrospinal fluid could be detected.

Asling *et al.* have demonstrated recently that the ataxia results from faulty embryonic development of the inner ear, particularly the otoliths of the utricular and saccular maculae.[28] These otoliths are characteristically absent or deformed in mice, rats, guinea pigs, and chicks[158] when there is a maternal deficiency of manganese. The defective otoliths do not show any metachromasia with toluidine blue staining,[452] which indicates that normal quantities of acid mucopolysaccharides are not present. Thus, it now appears that the newborn ataxia is attributable to the same basic biochemical defects that are responsible for the skeletal characteristics discussed earlier.

A similar ataxia is observed in mice with the mutant gene pallid;[322] in these mice, also, otoliths that do not stain for acid mucopolysaccharides are absent or defective. Moreover, in contrast with normal mice, but in common with manganese-deficient newborn, there is no uptake of sulfur-35 by the macular matrix in which the crystalline otoconia of the otoliths are embedded.[244] Erway et al. have shown that the histologic changes and ataxia can be completely prevented by adding manganese at 1,000 ppm to the maternal diet;[157] in other words, expression of the mutant gene is preventable. This is the first demonstration that a phenocopy-inducing agent—in this case manganese—can have a reciprocal effect on the gene itself. No other abnormal cartilage was detectable in the pallid strain, in contrast with observations in mice whose mothers have nutritional manganese deficiency.

Hurley has suggested that the local lesions in the inner ear may be secondary to lack of normal pigmentation of the membranous labyrinth in the pallid mice.[244] It has been reported by Cotzias et al.[114] that manganese is several times more abundant in pigmented tissue than in nonpigmented tissue, and it may be that the pigmented tissue is necessary to ensure an adequate local supply of manganese to the otoliths and that the effect of the absence of such pigmented tissue can be overcome only by high dosages of manganese.

GLUCOSE TOLERANCE

Everson and Shrader[162] have reported that oral-glucose tolerance is impaired in young adult manganese-deficient guinea pigs and could be corrected by dietary manganese supplementation. Fasting-blood-sugar content was increased in the deficient animals, with a mean 40 mg/100 ml higher than in controls. Peripheral utilization of parenterally administered glucose was also measured in cannulated nonanesthetized animals. The rate of disappearance of glucose from the blood was 0.8%/min in manganese-deficient guinea pigs, compared with 1.7%/min in control animals.

These investigators have also found that the newborn of guinea pigs that receive low intakes of manganese have abnormalities of the pancreas, with aplasia or hypoplasia of all cellular components. Those with hypoplasia had a reduced number of islets, and the islets contained fewer and less intensely granulated beta cells.[495] The young adult manganese-deficient animals with impairment of glucose tolerance also had a decreased number of beta cells and islets.

Pyruvate carboxylase was earlier found to be a manganese metalloenzyme,[445] with the manganese functioning in the transcarboxylation step,[352] but manganese is also involved in gluconeogenesis, in vitro confirmation of which has been obtained with studies using isolated perfused rat liver.[182] It is therefore apparent that manganese is important for normal carbohydrate metabolism and is required by both the pancreas and the liver.

REPRODUCTION

In the female rat, a relatively mild degree of manganese deficiency results in offspring with congenital malformations; a more severe degree of deficiency results in stillbirths and neonatal deaths; when there is an even greater degree of dietary manganese restriction, the animals will not mate or are sterile, with absent or irregular estrus cycles.[23] Similar findings have been reported in the guinea pig.[161] Manganese-deficient sows require more services per conception and farrow fewer piglets per litter than normal sows;[208] they also have irregular or absent estrus.[407] Cows have also been shown to have a disturbed estrus cycle when given a manganese-deficient diet.[48] Manganese supplements have improved fertility of cows under field conditions in different parts of Europe;[205, 231, 368, 535, 549] however, other factors may be involved, inasmuch as the reported dietary intakes of these cattle were higher than of others in which there was no manganese-responsive infertility (J. Hartmans, personal communication).

In the male rat or rabbit, manganese deficiency is associated with seminal tubular degeneration and absence of spermatozoa. These animals are sterile and lack libido.

The precise biochemical lesion responsible for these abnormalities of reproduction is unknown.

OXIDATIVE PHOSPHORYLATION

Lindberg and Ernster[306] demonstrated in 1954 that manganese is required as a cofactor for oxidative phosphorylation *in vitro;* more specifically, it is necessary for the coupling of phosphorylation to the oxidative reactions in the mitochrondria. Many other investigators have also reported effects of manganese on mitochondria *in vitro.* Maynard and Cotzias[334] have found that the uptake of manganese by mitochondria is higher than that by other cellular components.

Hurley[244] has reported that oxygen uptake by liver mitochondria from ataxic manganese-deficient mice is only 66% that of normal. How important the impairment of oxidation phosphorylation may be to manganese-deficient animals *in vivo* and whether it is linked to some of the observed features of manganese deficiency have not yet been clarified.

LIPID METABOLISM

In vivo, manganese deficiency results in increased back fat thickness in pigs,[407] which is reduced by later manganese supplementation. Manganese supplementation also reduces liver and bone fat of manganese-deficient rats.[9] Choline supplements produced similar results. The biochemical explanation of these observations and the nature of the lipotropic action of manganese are not known.

In vitro, manganese has been reported by Curran to be an activator in the synthesis of fatty acids,[121] and Curran has shown that the element stimulates the hepatic synthesis of cholesterol in rats. Manganese ion is a cofactor for the conversion of mevalonic acid to squalene.[8]

GROWTH

Manganese deficiency does not adversely affect growth to the same extent as does deficiency of some other essential trace elements, such as zinc. Growth retardation has been reported in manganese-deficient mice,[276] rats,[71] rabbits,[149] chicks,[187] and swine,[208] although other investigators[407] did not find reduced weight gain in swine. Low-manganese diets have not been associated with reduced weight gain in guinea pigs[161] or calves.[48]

BRAIN FUNCTION

Ataxia in the young of manganese-deficient mothers is not the result of brain dysfunction, as was initially thought; the metabolic defect in the inner ear is responsible for the ataxia, as discussed earlier.

However, there is limited evidence that manganese deficiency can result in some impairment of brain function. Manganese-deficient rats are more susceptible to convulsive states than normal rats.[249] This is independent of the presence of ataxia.

Biochemistry and Metabolic Role of Manganese

SUMMARY AND CONCLUSIONS

The specific biochemical role of manganese has been difficult to determine. Indeed, it has not been possible to specify a single biochemical role. However, whether it has one or many biochemical roles, its function in many cases cannot be taken over by other metals. Several important enzymes, such as prolidase and succinic dehydrogenase, demand manganese exclusively. Pyruvate decarboxylase is a mitochondrial manganese metalloenzyme that catalyzes the carboxylation of oxaloacetate from pyruvate. Manganese also plays a role in the function of other enzymes involved in carbon dioxide fixation, and arginase activity depends heavily on manganese.

It has been difficult, however, to relate specific manganese-dependent enzymes directly to gross pathology of manganese deficiency or excess. Only recently have enzyme systems been defined that are required for the synthesis of mucopolysaccharides and that provide a biochemical explanation for observed defects in connective tissue and skeleton. The same biochemical defects appear to be responsible for the ataxia that results from faulty embryonic development of the inner ear in the offspring of manganese-deficient animals. Manganese is important for normal carbohydrate metabolism, and a deficiency results in impaired glucose utilization and abnormalities of the pancreas. Manganese also is implicated in reproduction, lipid metabolism, growth, and brain function.

Manganese is an essential trace nutrient for microorganisms, plants, and animals, including all species of mammals and birds that have been investigated. Manganese deficiency has been observed in many mammalian species, both under field conditions and in the laboratory. It is therefore reasonable to conclude that man also has a nutritional requirement for manganese. The incidence of human manganese deficiency has not been investigated, nor has it been determined whether such a deficiency is a health hazard to man. Moreover, minimal human nutritional requirements have not been established. It will be necessary to determine such requirements if desirable limits of exposure to dietary and environmental manganese are to be established.

6

Manganese Toxicity and Catecholamines

Cotzias *et al.* have recently reviewed current knowledge of the metabolic modification of parkinsonism and of chronic manganese poisoning[112] and pointed out that there are apparent connections.

Much of the information on manganese toxicity is based primarily on clinical experience with humans subjected to prolonged exposure to manganese. Such clinical experience involves separation of symptoms due directly to intoxication and to patients' perception of symptoms. Interpretation of data from patients thus exposed is not easy. Experimental studies with animals, however, are few. Morphologic lesions of the central nervous system have been produced with manganese compounds. Real differences exist between species and maybe even within primates in susceptibility to these lesions. In at least some species, behavioral changes are evident, even if histologic alterations cannot be found. There is evidence that clinical and biochemical abnormalities may precede histologic changes and may be primarily biochemical—not anatomic.

This chapter deals with possible biochemical abnormalities, although it is admitted that there is very little direct information available on the biochemical and metabolic abnormalities that underlie the neurologic manifestations of chronic manganese toxicity. This paucity of data is attributable to a number of factors: metabolic studies of the brain are technically difficult, substantial information on the metabolic role of the biogenic amines in the normal brain has been acquired only recently and is still far from complete, and it has not been possible to use small laboratory animals as experimental models for the study of chronic manganese toxicity, because such animals as mice, rats, and guinea pigs are not susceptible to extrapyramidal neurologic disease.[115] Cotzias et al. have emphasized the correlation between the occurrence of extrapyramidal disease and the intensity of pigmentation in the substantia nigra;[115] these small animals have no visible pigmentation. In contrast, primates do have pigmentation of the substantia nigra and are susceptible to extrapyramidal disease, which can, for example, be induced by chronic exposure to toxic doses of manganese.[346] A recent biochemical investigation of the brains of squirrel monkeys after repeated injections of manganese dioxide has provided the first evidence of a specific metabolic abnormality in the brain attributable to chronic manganese toxicity.[374]

Direct investigations of the metabolism of the brain in patients suffering from manganese toxicity have not been reported. However, there are similarities between the clinical features of the extrapyramidal disease of manganism and that of parkinsonism. Moreover, Mena, Cotzias, and their colleagues[347] have demonstrated recently that these features in manganism respond favorably to therapy with L-dopa (levodopa) in a manner similar to that previously observed[111] in patients with parkinsonism. This similarity in therapeutic response provides strong indirect evidence that chronic manganism and parkinsonism share similar biochemical abnormalities with respect to the extrapyramidal system of the brain. It is therefore appropriate to summarize present knowledge on the metabolic abnormalities associated with parkinsonism.

The dopamine content of the basal ganglia is markedly reduced in patients with parkinsonism, compared with the normal.[53-55, 148, 239-241] Dopamine (3-hydroxytyramine), like epinephrine and norepinephrine, is a naturally occurring catecholamine in animal tissues. In chromaffin tissue and noradrenergic nerves, it is the immediate precursor of norepinephrine; however, it probably has a specific physiologic role of its own peripherally.[240] Approximately 50% of the total catecholamine content of the normal brain is present as dopamine; highest concentrations occur in the corpus striatum, substantia nigra, and globus pallidus of the extrapyramidal system. These same areas contain relatively small quantities of norepinephrine. It appears certain that dopamine has a specific neurophysiologic role, particularly in the extrapyramidal system, where it probably functions as a central neurotransmitter; however, the details of the role await final definition. Dopamine and brain function have been extensively reviewed by Hornykiewicz.[240]

Decreased dopamine content has been demonstrated in the corpus striatum, globus pallidus, and substantia nigra of patients suffering from idiopathic Parkinson's disease, postencephalitic parkinsonism, and parkinsonism attributable to reserpine.[240] They also have a reduction, although not nearly so severe, of 5-hydroxytryptamine and of norepinephrine in the hypothalamus. Other compounds, including acetylcholine and γ-aminobutyric acid, have also been implicated in these conditions. The decrease in dopamine in the brain of patients with parkinsonism attributable to a number of entirely different etiologic factors, but all with similar clinical features of extrapyramidal disease, strongly suggested a close relation between absence of or marked decrease in dopamine in the basal ganglia and these clinical features.

L-dopa is a precursor of dopamine that, unlike dopamine, can cross the blood–brain barrier; it is then at least partially converted to dopamine in the brain.[405] Cotzias and his colleagues have conclusively demonstrated beneficial effects of long-term, high-dosage L-dopa therapy in alleviating clinical features of parkinsonism.[111] However, the mere effectiveness of L-dopa therapy does not necessarily indicate that dopamine is of great importance. There is little doubt that L-dopa has beneficial effects in most cases of the disease, but this does not imply that it is a completely effective treatment—indeed, that is far from the case. The favorable response to L-dopa therapy, however, provides further evidence that a secondary depletion of dopamine may be important in the development of the clinical features of parkinsonism.

How dopamine is depleted in the brain is not yet clear. In catecholamine biosynthesis, the four enzymes primarily involved are tyrosine hydroxylase, aromatic amino acid decarboxylase, dopamine β-hydroxylase, and phenylethanolamine-N-methyltransferase.[358] However, phenylethanolamine-N-methyltransferase, as far as has been determined in most species, does not exist in the brain. The first step in the biosynthesis of L-dopa is the hydroxylation of L-tyrosine by tyrosine hydroxylase.[369,402,499] Tyrosine hydroxylase has been reported to be bound to synaptic vesicles in nerve endings where dopamine and norepinephrine also localize primarily their own neurons.[123,339] However, whether inactivity of this enzyme plays a role in the pathogenesis of parkinsonism is not yet known.[358]

Dopa decarboxylase is the aromatic amino acid decarboxylase that generates the corresponding amines from dopa and from 5-hydroxytryptophan.[462] The decarboxylation steps in the synthesis of the biogenic amines are all catalyzed by this same enzyme. However, it has been difficult to inhibit the enzyme sufficiently to decrease catecholamine content.[358]

Dopamine β-hydroxylase catalyzes the β-hydroxylation of dopamine to norepinephrine.[358] Endogenous inhibitors of dopamine β-hydroxylase are present in most tissues and act by complexing with the Cu^{+2} of the enzyme.

The final step in the biosynthesis of epinephrine is the N-methylation of norepinephrine by phenylethanolamine-N-methyltransferase, using the methyl group of methionine and S-adenosylmethionine. Phenylethanolamine-N-methyltransferase is strongly inhibited by its substrate, norepinephrine, and its product, epinephrine, and by heavy metals, such as Cd^{+2}, Hg^{+2}, and Cu^{+2}.[358]

Catecholamine content may be regulated at many sites, as well as the relative rates of synthesis and degradation of the four enzymes involved in catecholamine biosynthesis. Tyrosine hydroxylase is inhibited by such catecholamines as dopamine and norepinephrine, apparently because of competition between them and the pteridine cofactor required by the enzyme. Dopamine, normally found outside the vesicle with tyrosine hydroxylase, may be responsible for regulating the activity of tyrosine hydroxylase.[358] Each enzyme has specific cofactor requirements and differs in subcellular localizations within the cell. The available cofactors—such as tetrahydropteridine, pyridoxal phosphate, ascorbic acid, and S-adenosylmethionine—may also contribute to the regulation of catecholamine synthesis.

Although catecholamine biosynthesis primarily involves four enzymes, catecholamine metabolism involves two main enzymes: monoamine oxidase (MAO) and catechol-O-methyltransferase (COMT).[358] Each acts on a wide variety of amines and is fully active on the product of the other, which results in a spectrum of metabolites. About one third of the total catecholamine metabolites in human urine come from dopamine, the metabolites probably arising by the action of MAO on dopamine before it can be protected by uptake into storage granules. Storage of catecholamines (dopamine, epinephrine, and norepinephrine) in specific storage vesicles protects them from enzymatic destruction until released.[358] Reuptake and retention in the storage vesicles are the major mechanisms for inactivation.

Monoamine oxidase deaminates compounds in which the amine group is attached to the terminal carbon atom; but, because methylation and β-hydroxylation decrease the susceptibility to MAO, dopamine is metabolized more readily than epinephrine and norepinephrine.[358] Administration of MAO inhibitors may result in increased tissue contents of dopamine, norepinephrine, and serotonin,[358] depending on the tissue and the species. Dopamine can be readily deaminated oxidatively, but the isoenzyme of MAO mainly responsible for this reaction *in vivo* is not known.[112] The several species of MAO differ with respect to substrate specificity, heat stability, and effects of different inhibitors.[358]

L-dopa can be extensively O-methylated after administration.[493] The principal metabolite of L-dopa therapy is the O-methylated derivative. About one fourth of administered L-dopa is excreted in the urine as homovanillic acid.[30] Dopamine is the principal metabolite of endogenous L-dopa, and there is a striking increase, with passing months, in the fraction of administered L-dopa that is excreted as dopamine in the urine. The primate brain is rich in COMT.[31] S-adenosylmethionine is an essential cofactor in O-methyltransferase and N-methyltransferase reactions and becomes markedly diminished in animals receiving L-dopa.[553] High turnover in either reaction thus depletes the methionine pool on which the other also depends. Transmethylation reactions involving DNA and RNA, especially mRNA, have recently been detailed[24,199,555] and indicate how competition for S-adenosylmethionine can have far-reaching effects. Intervention of L-dopa into the transmethylation processes can deplete the methionine pool. With an average dose of L-dopa, the homovanillic acid excreted accounts for a large fraction of the methionine supplied by a regular American diet.[112] For S-adenosylmethionine to act as the transmethylation cofactor *in vitro* requires a divalent metal, such as manganese or magnesium.[32,67,86,87] The role of these metals in regulating transmethylation *in vivo* needs further evaluation.

A further metabolic abnormality typical of parkinsonism is a severe reduction in the melanin content of the nerve cells of the substantia nigra.[143] The precise physiologic role of melanin in the brain and the details of its relation to the metabolism of the biogenic amines are not known, but Cotzias *et al.* consider it likely to be of great physiologic importance to the brain.[115] Manganese also is normally associated with melanin.[114] On the basis of theoretical considerations of its chemical structure,[412] melanin is concluded to be a semiconductor; semiconductors may well be very important in the central nervous system. Only primates, man, and some of the other large mammals have pigmentation of the substantia nigra; and only in these same species has extrapyramidal neurologic disease been observed. Two such conditions, parkinsonism and phenylpyruvic oligophrenia, are known to have reduction or absence of melanin in the basal ganglia.[115] Again, the precise connection between the reduction of melanin, the reduction in dopamine, and the clinical features is not yet entirely clear. That melanin and catecholamine metabolism are related is, however, further suggested by the following data:

1. Melanocytes and sympathetic nerve cells are both derived from the neural crest.
2. In both melanocytes and sympathetic nerve cells, tyrosine is hydroxylated to dihydroxyphenylalanine (dopa).
3. Dopa is a precursor for both melanin and catecholamines.
4. In the brain, but not in the skin, dopamine appears to be a main precursor of melanin.[534]
5. Melanin granules contain substantial quantities of MAO.

In summary, then, the two major biochemical abnormalities known to be associated with parkinsonism are a depletion of brain catecholamines—especially dopamine in the substantia nigra, globus pallidus, and corpus striatum—and a depletion of melanin granules in the substantia nigra. There is now considerable evidence that the melanin granules are intimately linked with the metabolism of the catecholamines. However, whether depletion of melanin precedes or is secondary to the impairment of catecholamine metabolism is not known; nor is it understood how melanin contributes to the normal neurophysiology of the brain, especially the extrapyramidal system.

Although the precise physiologic role of the melanocytes in the brain is not understood, they merit consideration not only because of their depletion in parkinsonism, but also because of the apparent importance of manganese in the metabolism of melanin granules.[107]

Manganese is normally concentrated in melanin.[114] Cotzias *et al.* have postulated that, although this normal concentration is required for optimal function of the melanocytes, additional accumulations resulting from chronic manganese toxicity could be deleterious.[115] Melanin granules also contain semiquinone free radicals. *In vitro,* interactions between manganese and phenothiazines yield semiquinone free radicals.[64,65] The phenothiazine drugs—which, like manganese, accumulate in melanin granules—can also cause a parkinsonian syndrome in man, possibly by blocking the normal metabolism of brain dopamine, rather than by decreasing the brain content of this biogenic amine.[240] At present, these remain isolated observations; but they at least suggest the possibility that the known interrelation between parkinsonism, melanin granules, phenothiazines, and manganese is more than coincidental. Although one can only speculate on the possible importance of melanin in both the normal brain and the brain of a person with chronic manganese poisoning, this subject certainly merits further research.

The clinical similarities between chronic manganese poisoning and parkinsonism and, above all, the excellent clinical response of patients with chronic manganism to therapy with L-dopa[347] indicate that abnormalities of brain catecholamine metabolism are a cardinal feature of the disease. However, direct evidence of this is limited.

Cotzias and colleagues[109,334] have demonstrated that manganese and MAO were both concentrated in mitochondria and suspected that their metabolism might be coupled.[108] The interrelations between manganese and melanin granules, on the one hand, and melanin granules and biogenic amines, on the other, have been discussed earlier. In 1968, Papavasiliou and colleagues reported the finding of a direct link between the biogenic amines and manganese.[390] The liver was used for these studies because of the enormous complexities of direct investigations of the brain. L-dopa, dopamine, L-epinephrine, and DL-isoproterenol—but not tyramine or norepinephrine—were found to increase the intracellular concentration of manganese. Glucagon and cyclic $3',5'$-adenosine monophosphate ($3',5'$-AMP) duplicated these results, and it was concluded that the effects of the biogenic amines on manganese metabolism were mediated through the increase in intracellular cyclic $3',5'$-AMP. Sutherland and Robison[470] have correlated the potency of different amines with the capacity of each to increase the production of cyclic $3',5'$-AMP. Neither the importance of Papavasiliou's observation to normal brain metabolism nor the possible effect of excess manganese on these findings is known. The study of Papavasiliou et al.[390] was the first to demonstrate a definite connection between the metabolism of manganese and catecholamines.

Direct evidence of an effect of manganese toxicity on brain biogenic amines has come recently from the study by Neff and colleagues[374] using squirrel monkeys. The brains of these monkeys were studied histologically and chemically after repeated subcutaneous injections of 200 mg of manganese dioxide. The concentration of dopamine was significantly reduced in the caudate nucleus within 4 months of the initial injection. Moreover, the extent to which the dopamine content was reduced correlated well with the degree of toxicity as assessed by observed neurologic (extrapyramidal) abnormalities. Norepinephrine concentrations in the cerebrum and brain stem were not reduced; serotonin concentrations were normal in the cerebrum but markedly reduced in the caudate nucleus.

It is generally believed that lower brain dopamine content in parkinsonism results from the destruction of "dopaminergic" neurons. It is therefore of particular interest that histologic examination of the brains, including the basal ganglia, of the monkeys suffering from manganese toxicity did not reveal any histopathologic changes. It appears that clinical and biochemical abnormalities may be evident before there is any macroscopic destruction of nerve cells. In human manganism, this may be particularly relevant to the early psychotic manifestations, which could also be attributable to a disturbance of amine metabolism.[348]

The animal studies have provided very strong evidence that the metabolic abnormalities in the brain caused by chronic manganese toxicity, particularly the decrease in dopamine content, are very similar to those occurring in parkinsonism. This similarity is also strongly suggested by the similar favorable response of both human diseases to L-dopa therapy.[111,347] Although the biochemical nature of the response to L-dopa requires further investigation,[552] there is already evidence[405] that L-dopa does, indeed, increase brain dopamine content. The unfavorable response to L-dopa therapy in one case of manganism, in which hypotonia was a prominent feature, and the later improvement after administration of the serotonin precursor 5-hydroxytryptophan indicate that a deficiency of brain serotonin may be more important in some cases.[347] It is likely that the balance between dopaminergic and tryptaminergic receptors is important for normal extrapyramidal function.[298]

In summary, there are close similarities between the clinical features of the extrapyramidal disease of manganism and those of parkinsonism. They provide strong indirect evidence that chronic manganism and parkinsonism share similar biochemical abnormalities with respect to the extrapyramidal system of the brain.

The dopamine content of the basal ganglia is markedly reduced in patients with parkinsonism. L-dopa is a precursor of dopamine, which, unlike dopamine, can cross the blood–brain barrier, where it is converted to dopamine in the brain. Long-term, high-dosage L-dopa therapy has beneficial effects in most cases of parkinsonism, and there is excellent clinical response to therapy with L-dopa in patients with chronic manganism. Although the biochemical nature of the response to L-dopa requires further investigation and L-dopa is not a completely effective treatment, favorable response to L-dopa therapy in both conditions indicates that a secondary depletion of dopamine may be important in both. The brain dopamine concentration has been shown to be markedly reduced experimentally in monkeys by manganese injections, and there is evidence that L-dopa increases brain dopamine content in man.

Also typical of parkinsonism is a severe reduction in the melanin content of the substantia nigra. The melanin granules appear to be intimately linked with the metabolism of the catecholamines. Manganese also is normally concentrated in melanin, and the known interrelation between parkinsonism, melanin granules, and manganese may be more than coincidental. The importance of melanin in the normal brain and in brains of patients with manganism merits further research.

Animal studies and comparisons of acute and chronic manganese poisoning with parkinsonism, together with observed effects of L-dopa therapy, indicate that a major abnormality may be a defect in the metabolism of dopamine.

7

Epidemiology of Manganese Intoxication

NEUROLOGIC MANIFESTATIONS OF MANGANISM

Until 1837, manganese metal was considered to be inoffensive to human health. At that time, Couper,[118] on the basis of disorders displayed by five workers of a pyrolusite mill, presented a clinical description of manganism and expressed the opinion that manganese can provoke a condition similar to paralysis agitans. These observations fell into oblivion.

In 1901, Jaksch[258] and Emden[152] rediscovered the disease and described three and four cases, respectively.

According to the statistics of Voss,[525] there were 152 cases of manganism known in the medical literature until 1935 and the sources of the disease were the manganese ore mills, the voltaic-cell factories, and the iron and steel industry. Until then, no one had recorded the existence of the disease in manganese miners. In 1879, Schlockow[439] had described 40–50 cases of walking disorders in zinc miners of Upper Schleswig—"a peculiar disorder of the spinal cord in zinc workers." It is now believed that those miners suffered from manganism provoked by the inhalation of manganese dust, which was present in high concentrations in the zinc ore, or by the inhalation of manganese vapors during the smelting of zinc.

In the 1930's and 1940's, Dantin Gallego[125,126] described two cases of manganism in manganese miners of Huelva, Spain. In 1936, Nazif[372] and Scander and Sallam[438] described 26 cases of manganism in miners of the Sinai; these were confirmed by Baader[33,34] during his study trip to Africa in the same year, during which he also detected manganism in Moroccan miners.

In 1936, Büttner and Lenz[84] examined a group of miners who had worked in the manganese mines of Giessen, Germany, for more than 20 years and found 11 mild cases and one severe case of manganism. In the same group of miners, they found 38 cases of manganosilicosis.

In the following years, owing to the necessities of the war economy and the postwar development of the iron and steel industry, the extraction of manganese was stepped up. The extension of pneumatic drilling in mining work led to an increase in the number of miners. Mining work was the main cause of the growing incidence of chronic manganese intoxication, which was sufficient to take the condition out of the category of medical rarities.

In 1944, Ansola et al.[21,22] described 64 cases of manganism in miners of Antofagasta, Chile, and noted an additional 11 probable cases. A special aspect of the clinical material studied by these authors was the association of the extrapyramidal syndrome and psychic disorders (psychomotor hyperkinesia, impulsiveness, and euphoria).

Schuler et al.[444] showed that the introduction of pneumatic drilling, with an increase in dustiness, had led to such an outbreak of manganism at the Coquimbo, Chile, manganese mines in 1953 that, after 3 months of the new procedure, the mine returned to manual drilling. Wet drilling was prevented by a water shortage in the region. Fifteen cases of manganism were discovered among 83 miners who had worked in this mine for a long period.

In a survey of the Moroccan manganese mines in 1947, Rodier and Rodier[429] found 28 typical cases of manganism among 257 miners. Between 1951 and 1958, 223 cases of manganism were diagnosed in the same mine; of these, 158 were in pneumatic-drill operators.[426] In 1958, Ritter[421] described nine cases of manganism in the Idikel-tafraout mine in Morocco that occurred after exposure periods of only 6–8 months.

In Cuba, the studies by Garcia Avila and Penalver Ballina[188] in 1952 led to the detection of 120 cases of manganism in the Charco Redondo mine. In later investigations, they detected other cases of manganism.[396]

In Italy, Prosperi and Barsi[411] examined the morbidity of miners in a pyrolusite mine in Tuscany and did not find a single case of manganism.

They attributed the absence of manganism to the high water content of the ore deposits, which reduces the risk of dust inhalation.

In Mexico, Roldan[431] carried out a survey of 151 miners and discovered 12 cases of manganism (8%) in workers 22-35 years old who had been working 12-16 h a day for 1-2 years. The clinical picture was characterized by a combination of psychomotor excitation phenomena and extrapyramidal disorders typical of manganism.

In Japan, Suzuki *et al.*[472] examined 237 manganese miners of the Shikoku region in 1950-1959. They noted that 30% of the miners suffered from shaking of the trunk and limbs, micrography, disturbances of walking or motor coordination, and especially subjective disorders. They also noted progressive worsening of the clinical picture during the period of observation.

The USSR studies of the Ciatursk manganese miners by Saradzhashvili (cited in Khavtasi[280]) in the 1940's did not reveal cases of manganism. However, Khavtasi later examined 972 miners of these mines and found 69 of them suffering from manganism.[280]

At Nicopol, USSR, Khazan *et al.*[281] examined 170 miners in 1956 (75% of whom had worked there for more than 10 years) and found 39 suffering from central nervous system diseases (ranging from the "asthenovegetative" to the extrapyramidal syndrome) and 40 with diseases of the peripheral nervous system (especially sacrolumbar radiculitis). In the same group of 170 miners, they detected 21 cases of manganoconiosis in various stages.

In Rumania, studies of 576 miners of Suceava in 1949 and 1950 revealed isolated or associated neurologic symptoms in 53% and 10 cases of manifest manganism.[532]

Flinn *et al.*[171] reported a study of a Pennsylvania manganese-ore crushing and packaging mill in 1939 in which 11 of 34 workers suffered neurologic symptoms of manganese poisoning. There were few hygienic precautions, and the mill was extremely dusty. Cases stopped occurring when the mill was relocated and modernized and a medical examination program was initiated. Over the years, the hygienic measures were gradually abandoned; and new cases were reported by Tanaka and Lieben[479] in 1969. Even after that, industrial hygiene recommendations were not complied with, and four more cases occurred between 1969 and 1971 (S. Tanaka, personal communication). No cases of pneumonia have been reported from the mill, but this may be only because they were not looked for.

Whitlock *et al.*[538] reported two cases from a small manganese-steel operation, in which castings that contained 10-12% manganese received their final preparation by having rough spots burned off with a welder's torch; manganese poisoning was produced by the inhalation of fumes that contained relatively small amounts of manganese.

Schuler et al.[444] have described in detail the neurologic symptoms of manganese poisoning, as follows:

> Selections for examination were based upon histories of high exposure or upon spontaneous complaints of symptoms generally associated with manganese poisoning. Fifteen workers [with] typical manganese poisoning [were] analyzed ... for the purpose of contributing to a clear understanding of this clinical syndrome.
>
> The exposure periods [of 15 miners] ranged from nine months to 16 years, with an average of about eight years. [Ages ranged from 21 to about 50 (Table 7-1). The symptoms in several of the cases started] during a brief period of pneumatic drilling.... 13 [cases] occurred among drillers and two in samplers.
>
> *Associated Health Problems.* Alcohol was consumed sparingly or not at all by 11 of the miners ... and was used moderately or excessively by four. [It was] concluded that previous or concomittant [sic] alcoholism is not a requirement for the development of industrial manganism. No relationship was discovered between development of the condition and personal or familial histories of psychiatric disturbance. In one [case] the typical psychiatric disturbance of manganism had its onset during an attack of hepatitis; but acute disease, including infectious processes, seemed to play no part in the onset of any other case. [The psychomotor disturbances and neurologic symptoms and signs of the 15 miners are summarized in Tables 7-2 through 7-4.]

TABLE [7-1] Distributions of Ages and Exposure Periods of 15 Miners with Manganese Poisoning

A. Age distribution	
Years	No.
21–30	2
31–40	6
41–50	5
Over 50	2
B. Exposure periods	
Years	No.
Under 1	1
1–3	2
4–6	2
7–9	3
Over 9	7
Average	8 years, 2 months
Minimum	9 months
Maximum	16 years

Epidemiology of Manganese Intoxication

TABLE [7-2]   Psychomotor Disturbances in 15 Cases of Manganism

Type of Disturbance	Disturbances in Each Case	Cases in Which Each Disturbance Occurred	
		Number	Percent
Emotional instability		8	53
Irritability	+ + + − + + + + − − − − − − −	7	
Restlessness	+ + − + + − − − − − − − − − −	4	
Tendency to weep	+ + − − + − − − − − − − − − −	3	
Withdrawal from group	+ + − − − − + − − − − − − − −	3	
Unmotivated laughter	− + − − − − − − − − − − − − −	1	
Apathy	+ − + − − + − + + + + − − − −	7	47
Hallucinations	+ − + − − − − − + + − + + − −	6	40
Flight of ideas	+ − + + − + − − − − − − − + −	5	33
Compulsory acts	+ − − + + − − − − − + + − − −	5	33
Verbosity	+ + + + − − − − − − − − − − −	4	27
Total number disturbances in each case	9 6 5 4 4 3 2 2 2 2 2 2 1 1 0		

Comment

The observations in this study support the generally held view[22, 70, 427, 428]* that inhalation represents the principal hazard in industrial exposures to manganese. This point was adequately demonstrated at the mine responsible for the present cases when the incidence rate of new cases rose sharply during a brief period of dry pneumatic drilling. Unquestionably, intestinal absorption must occur since part of the inhaled dust is subsequently coughed up and swallowed, and the drinking water available at the place of work is frequently contaminated with manganese dust. From the observations of Flinn [et al.][171] the lungs may be expected to show manganese concentrations eight times that found in the liver and intestines; while the kidneys accumulate still less, and the brain . . . shows the lowest concentration. . . .

Most of the manganese reaching the liver is excreted in the bile and eliminated in the feces. A lesser proportion is subsequently removed from the blood stream by the kidneys and is eliminated in the urine. . . .

That airborne and inhaled manganese dust represents the principal hazard is established by the observations that the time of onset of manganism is inversely related to the concentration in air of the agent, and that, in rough terms, the severity of the disease is apt to be directly related to this concentration. In Ansola's [et al.] experience[22] the pre-symptomatic exposure periods averaged six months, and in Rodier's[427] the average was less than two years with a range of one month to 10 years. The average . . . exposure time . . . in [these] cases was about eight years with a range of nine months to 16 years. . . .

*Reference numbers have been changed to conform with this volume.

TABLE [7-3] Neurological Symptoms in 15 Cases of Manganism

Symptom	Symptoms in Each Case	Cases in Which Each Symptom Occurred	
		Number	Percent
Asthenia and adynamia	+ + + + + + + + + + + + − + +	14	93
Sialorrhea	+ + + + + + + + + + + − + + −	13	87
Fatigability	+ + + + + + + + − + + − + + +	13	87
Cephalalgia	+ + + + + + + + + + − + + − −	12	80
Disturbances of sleep	+ + + + + + + + + + − + + − −	12	80
Muscle pains	+ + − + + + + − + − + + − − +	10	67
Paresthesias	+ + + − + + + + + − + + − − −	9	60
Diaphoresis	+ + + + + − − − + − − − − − −	6	40
Disturbances of speech	+ − + + − + − + − − − − − − −	5	33
Disturbances of libido	+ + + + − − − − − − − − − − −	4	27
Disturbances of ejaculation	+ + + − − − − − − − − − − − −	3	20
Total number symptoms in each case	11 10 10 9 8 8 7 7 7 5 5 4 4 3 3		

TABLE [7-4] Neurological Signs in 15 Cases of Manganism

Signs	Signs in Each Case	Cases in Which Each Sign Occurred	
		Number	Percent
Muscular hypertonia	+ + + + + + + + + + + + − − +	13	87
Expressionless face	+ + + + + + + + − + − − + − −	10	67
Gait changes	+ + + + + + + + − + − − + − −	10	67
Monotonous voice	+ − + + + − + + + − − − − − −	7	47
Tremor, upper extremities	+ + + − − + − − + − − − − + −	7	47
Tremor, lower extremities	+ + + + − + − − + − + − − − −	6	40
Superficial sensory disturbance	+ + + + − − − − − − − − − − −	4	27
Deep sensory disturbance	+ + + − + − − − − − − − − − −	4	27
Impaired hearing	− − − + − − − − − − + − + − −	3	20
Postural changes	+ + − − − − − − − − − − − − −	2	13
Diplopia	− + − − − + − − − − − − − − −	2	13
Pyramidal signs	− − − − + − + − − − − − − − −	2	13
Total number signs in each case	9 9 8 7 6 6 5 4 4 3 2 2 2 2 1		

Epidemiology of Manganese Intoxication

The variability of time exposure required . . . to develop the disease may be explained on the basis of frequent changes in jobs and in the material mined, as well as by variations in the natural ventilation which in turn resulted in fluctuations in concentrations of airborne manganese dust. . . .

. . . Individual susceptibility, although possible, was not evident in the present cases, nor could differences in toxicity among different ores be discerned. There has been some consideration that latent manganism may become clinically manifest in the presence of an infectious disease. This . . . occurred in only one [case] of infectious hepatitis. The association of pneumonia, suggested in the literature of other countries, was not noted in our experience.

Summary

1. This study was made at a manganese mine whose main product is pyrolusite (MnO_2). The mining methods in use were primitive, natural ventilation was poor, and mechanical ventilation was unavailable; all of [this] resulted in high manganese exposures.

2. Of 39 air samples taken, four were in the range of the M.A.C. (6 mg. Mn/M^3 air), 22 were below, and 13 were above this value. The wide variations in air concentrations are due to changes in the nature of the ore encountered and to changes in the natural ventilation. Although the data do not permit a [reasonable] estimation of the integrated dose of any person studied, it may be stated that the exposure of these miners frequently exceeded the M.A.C. value. Drillers had the heaviest exposures.

Stokinger[466*] describes manganese poisoning as follows:

The onset is insidious, with apathy, anorexia, and asthenia. The Mn psychosis . . . has certain definitive features: unaccountable laughter, euphoria, impulsiveness, and insomnia, followed by overpowering somnolence. Headache is often present; recurring leg cramps; and sexual excitement, followed by impotence. Following or concomitantly with these manifestations are speech disturbances with slow and difficult articulation, incoherence, even complete muteness. Mask-like facies sets in, general clumsiness of movement, noticeable in altered gait and balance, which may develop into severe propulsive and retropulsive movements, and ultimate development of 'hen's gait.' Micrographia is a consistent finding. As the poisoning progresses, rigidity is marked, frequent falls occur, and tremor sets in, which becomes exaggerated by stresses such as fatigue or emotion. Absolute detachment, broken by sporadic and spasmodic laughter, ensues and, as in extrapyramidal affections, salivation and excessive sweating. Despite the severe incapacitations imposed by the disease, the patient survives, although permanently disabled unless treated; chronic Mn poisoning is not a fatal disease.

. . . Factors possibly influencing sensitivity are alcoholism, chronic infections, such as syphilis, malaria, tuberculosis, and avitaminosis, and liver dysfunction. The

*Reprinted with permission of John Wiley & Sons, Inc., copyright © 1962.

difference in time of onset and severity of the chronic disease seen throughout the world, although recognizably influenced by all these factors, still might be most importantly influenced by nature of the Mn ore; Rodier has stated that braunite, a mixture of Mn_2O_3 and $MnSiO_3$, is much more injurious than pyrolusite (MnO_2).

No indisputable test of Mn poisoning has been developed, but Rodier[427]* finds diminished excretion of 17-ketosteroids in 81 percent of patients, a relative increase in lymphocytes, and a decrease in the number of polymorphonuclear cells in 52 per cent, and an increased basal metabolic rate. Increased urinary Mn is evidence of exposure, but the correlation with symptoms is not clearly established. Kesíc and Häusler[278] found slightly higher (4.5×10^6 vs. 4.337×10^6) erythrocyte counts and decreased monocyte values (640 vs. 780) among Jugoslavian miners during the first phase of the disease, that later returned to normal; in only a few cases (4 of 60) was there a decreased leucocyte count. Fairhall [and Neal], in [their] review,[163] [mention] early rises in red blood cell count as the first symptom of the disease. At the present time, blood Mn levels are not used; Flinn et al.[170,171] found no Mn in the blood of the specimens they examined with the methods of the time (volumetric bismuthate); Penalver[395,396] states, without presenting evidence, that blood Mn levels are too variable to be of value. Possibly, application of newer analytical methods would prove of value.

The characteristic pathological lesion in man is destruction of the ganglion cells of the basal ganglia, followed by scarring and shrinking. Perivascular degenerative areas in the striatum and pallidum occur and to lesser extent in the frontal and parietal cortex. Others[525] believe Mn effects are not confined to the corpus striatum but involve the vegetative and vasomotor centers of the midbrain and also the medulla and cord. Unfortunately, owing to the customs in the countries in which manganism is currently most prevalent, autopsies cannot be performed and thus no recent histological examinations have been reported.

The prognosis of manganism depends on the duration of the disease; many of the symptoms will regress or disappear if the worker is removed from exposure shortly after the appearance of symptoms, although there may be some residual disturbances in speech and gait. Well-established manganism, however, is a crippling disease with permanent disability particularly in respect to the use of the legs.... Some of these patients may improve slowly with occasional remissions of tremor, weakness, and muscular cramps. Seriously poisoned individuals have been considered lifelong cripples. Recently, however, Penalver[395] has reported marked improvement in a chronic case of Mn poisoning on prolonged treatment with EDTA [ethylenediaminetetraacetic acid] salt plus other drugs offering supportive treatment. Prior work with EDTA in animals[35,291] given single or repeated doses of Mn salts had shown EDTA effective in removing recent Mn from body tissues, possibly preventing the disease from advancing, but not in curing the nervous effects of Mn.

Mobilization of manganese in manganism by EDTA has been reported

*Reference numbers have been changed to conform with this volume.

elsewhere. Urinary concentrations were said to increase fivefold. Some cases with mild symptoms improved considerably; others showed no changes.

Stokinger continues:

> Reports of more human cases treated with EDTA would be welcome. In a comparative test of three EDTA-type compounds against acute experimental Mn poisoning in rats, diethylene triaminepentaacetic acid was found to be somewhat more effective in preventing Mn intoxication than EDTA.[181]

MANGANESE INTOXICATION BY WELL WATER

The neurologic manifestations of manganese poisoning appear to be caused primarily by inhalation of dust or fumes; some consider ingestion as an additional factor. A waterborne epidemic has been reported by Kawamura *et al.*:[275] An encephalitis-like disease occurred in six members of a family. All had the same symptoms, including loss of appetite, constipation, and a mask-like facial expression, with running saliva. Tonicity of muscles was decreased; the leg joints were painful and rigid; the arm muscles showed rigidity and tremors; there was temporary double vision; tendon reflexes were increased; and there was some mental disturbance, memory loss, and melancholia. Preliminary investigation determined that one victim had died, two were hospitalized, and three were up and about. Blood and spinal-fluid samples were sterile, with normal cellular counts. Histologic examination of the autopsy material from brain and spinal cord showed no signs of encephalitis, such as perivascular cellular infiltrations, changes of nerve cells, increase of glia cells, and neuronophagy.

Clinical pictures of the cases and laboratory examinations of the materials indicated that the epidemic was not encephalitis. The symptoms pointed strongly to some form of intoxication. Carbon monoxide was discounted. It was soon learned that the family maintained a bicycle-repair shop and that many old dry cells for bicycle lamps had been buried near a well that supplied water for the family. It was presumed that the intoxication was caused by manganese, which, with zinc, is a principal constituent of the cells. This substance dissolved and contaminated the well water. Chemical analysis showed unusually high concentrations of manganese and zinc in the water from this and several other wells. Manganese and zinc were found in large quantities in the viscera of the autopsied victims and in the blood and urine of survivors. Ten more patients were discovered among the neighbors of the family; all had drunk contaminated water.

MANGANESE PNEUMONIA AND PNEUMONITIS

Brezina,[74] in 1921, was the first to report the relation between manganese dust and pneumonia. He reported that five of 10 exposed workers in the Italian pyrolusite industry died of pneumonia after 6 months of exposure in the course of 2 years.

In 1939, Elstad[150] reported the occurrence of lobar pneumonia in the Norwegian town of Sauda. A plant began smelting ferromanganese there in 1923. Toward the end of that year, six inhabitants died of lobar pneumonia; since then, the disease has claimed steadily increasing numbers of victims. Although mortality from all causes was practically identical for Sauda and the rest of Norway, the pneumonia mortality was eight times as high in Sauda (3.27 per 1,000 inhabitants) as in the rest of the country (0.4 per 1,000). Between 1924 and 1935, lobar pneumonia accounted for 3.65% of all deaths in all of Norway and 32.30% of all deaths in Sauda. The disease had been infrequent in the community until the operation of the plant.

The plant's electric furnaces produced ferromanganese containing 80% manganese and silicomanganese. Manganese-containing smoke was allowed to escape from the buildings housing the furnaces, and the smoke was a source of air pollution for the inhabitants of Sauda. The town is in a fjord surrounded by mountains. The dry matter in the smoke contained 54% silica and 2.56% manganese oxide near the plant. Farther away, the silica concentration increased and the manganese concentration decreased. Both manganese and silica particles occurred in sizes of less than 5 μm.

Over the years, the numbers of pneumonia cases and deaths varied with the tonnage of manganese alloy produced. Lobar pneumonia in Sauda attacked not only workers of the plant but all inhabitants of the community. It was also highly communicable, and multiple cases occurred in some homes.

Lloyd Davies[311] studied the incidence of manganese pneumonia in the manufacture of potassium permanganate. Men exposed to the inhalation of manganese dioxide dust and higher oxides of manganese suffered a high incidence of pulmonary symptoms, in particular illness diagnosed as pneumonia. For 1938–1945, the incidence of pneumonia was 26/1,000 among the men so exposed; that in an unexposed but otherwise similar group was 0.73/1,000. In the manufacture of potassium permanganate, manganese dioxide is ground, reacted with potassium hydroxide in the presence of lime, roasted in rotary kilns, and submitted to electrolysis so that potassium manganite, manganate, and permanganate are formed in succession. Potassium permanganate and manganate spray is liberated in the electrolytic process. During the reactions, a thick cloud of steam and dust—mostly manganese dioxide and lime— is given off; the plant was always very dusty. Dustiness varied with the weather. Dust samples were collected with a thermal precipitator and ranged from fewer than 1,000 to 37,000 particles/cc; 80% of the particles were smaller than 0.2 μm, and nearly all were smaller than 1 μm. The calculated mass concentration was 0.7–38.3 mg/m^3. The manganese content (expressed as manganese dioxide) of the dust collected with an electrostatic precipitator varied from 41 to 66%.

Other cases of manganese pneumonia and pneumonitis have been reported by authors of many different nationalities.[33,35,81,84,180,221,372] Most of the reports of chest symptoms, pneumonia, and pneumonitis refer to statistical studies or to single cases. The term "manganosilicosis" is used by several authors and obviously refers to a combined disease in miners exposed to both silica and manganese, either together or in succession.[122] Few details are mentioned. No reports of pneumonia or pneumonitis in manganese workers have emanated from American authors. The reports stress that manganese may potentiate or aggravate infectious processes, producing increased frequency, severity, and mortality.

Schopper[441] described the postmortem appearances of two miners who had died of pneumonia. In addition to the ordinary signs of pneumonia, he found throughout the lungs extensive deposits of foreign material rich in manganese, which had led to fibrosis of the lung.

A type of pneumonia similar to that reported in manganese workers has been reported in basic-slag workers.[269] It is considered an occupational disease related to the processing, bagging, and loading of Thomas slag* or Thomas meal, which contains 6–8% manganese. A relatively large fraction of exposed persons suffer from pneumonia, and their case mortality is 20–30 times higher than in ordinary pneumonia. Baader[35] assumes that there is great similarity between manganese and

---

*Thomas slag is finely powdered basic slag, a fused product that separates in metal smelting and floats on the bath of metal; it is formed by the combination of flux with gangue of ore, ash of fuel, and perhaps furnace lining.[216] It is obtained in the Thomas process of making steel, using burned dolomite, which reacts with phosphoroys of the pig iron to form a converter lining. It consists of phosphates, and it is used as a fertilizer.

Thomas-slag pneumonia and that the chest symptoms are caused primarily by the manganese in the slag.

DIAGNOSIS

Like many other occupational diseases, manganese poisoning has a symptomatology similar to that of illnesses that occur in the nonexposed population—e.g., diseases of the basal ganglia, such as parkinsonism and pneumonia.

The diagnosis of manganese poisoning is based on symptomatology; proof of overexposure by findings of increased manganese in urine, blood, and hair; statistical evidence; and exclusion of similar diseases. Individual susceptibility plays an important role. There are always workers who do not develop symptoms even though they were exposed to the same concentrations as those who fall ill.

The development of the disease is related to the manganese concentration in the air, the duration of exposure, and the size of the manganese particles in the air.

In manganism, there is usually a low white-cell count and an increase in blood manganese content. Urinary manganese content is a more reliable sign of exposure for use in the diagnosis. Manganese is increased in hair and fingernail clippings in manganism.

Multiple sclerosis, paralysis agitans, and cerebrovascular syphilis must be considered.[163] Negative blood tests for syphilis and characteristic colloidal curves of the cerebrospinal fluid for multiple sclerosis may help in the differentiation. Removal from exposure to manganese will stop the progression of manganism but not of the other diseases.

Manganism may be mistaken for epidemic encephalitis but may be distinguished from it by the absence of pathologic changes in the eyegrounds.[351,521]

Another illness that can be mistaken for manganese poisoning is progressive lenticular degeneration (Wilson's disease), which is associated with a degeneration of the liver.[465]

SUMMARY

Until 1837, manganese was not incriminated as a disease producer. After an initial report of a neurologic disorder closely resembling parkinsonism by Couper, reports of a similar disease appeared from many parts of the world. The disease occurred in manganese miners (ore-crushing-plant workers), steel workers, chemical workers, and others who had the opportunity to inhale manganese dust or fumes. A form of pneumonia or pneumonitis was reported by Elstad in inhabitants of a Norwegian town contaminated by the effluent of a metallurgic plant producing ferromanganese. This was followed by reports of a similar disease from many parts of the world. Inhalation of manganese is generally more likely than ingestion to be a route of entry for the production of the disease. Japanese authors reported an outbreak of manganese poisoning that resulted from ingestion of manganese-contaminated well water.

8

Permissible Air Concentrations of Manganese

The threshold limit value (a ceiling value) for manganese recommended by the American Conference of Governmental Industrial Hygienists is 5 mg/m^3, which is generally believed to carry a low margin of safety for those occupationally exposed. According to Public Health Bulletin 247,[171] the lowest average concentration at which a case of chronic manganese poisoning was found was 30 mg/m^3. This was in a manganese mill; the worker was exposed to manganese dioxide dust. Manganese fumes may be more toxic, inasmuch as lower concentrations of fumes have produced cases of chronic manganese poisoning in several steel plants.[479,538]

As with many other substances, there is a wide range in the amounts of manganese to which workers in various countries are permitted to be exposed. These range from 0.3 mg/m^3 in the USSR, Poland, and Hungary to a peak concentration of 10 mg/m^3 in Czechoslovakia.[122] The United States, with a ceiling value of 5 mg/m^3, stands in the middle, closely followed by East Germany (the German Democratic Republic), West Germany (the Federal Republic of Germany), and Yugoslavia, with 6 mg/m^3.

The wide range is based on different concepts and interpretations. The USSR and some of its followers depend heavily on behavioral changes and reflex modifications as bases for their low target standards.

The United States and other countries base their standards, which are enforced, on air sampling in the working environment and absence of obvious disease at specific concentrations.[10] This difference arises out of basic differences in the definitions of a state of health and out of conflicting views concerning the degree to which man's physiologic defense mechanisms can be safely drawn on to offset the effects of offending agents.[216]

In the United States, dose–response studies have been conducted, ranging from the greater exposures resulting in demonstrable ill effects down to small exposures at which increasingly sensitive measures of preclinical, physiologic, biochemical, and other indices of functional disturbances are required to demonstrate effects. However, these new measures have been constantly subjected to critical tests of usefulness as predictors of ill health. Threshold limit values are set to ensure that the kind and degree of response are kept below the limits of significance. In the USSR, in addition to physiologic and biochemical studies involving preclinical conditions, emphasis is put on starting at the other end of the dose–response relation, and, by working up from zero dose and an initial benchmark of normality in the test organism, the permissible limit of exposure is established below the lowest dose needed to induce a statistically significant departure from the normal state, as revealed by highly sensitive measures of behavioral response. The differences in concept of the healthy state are clear: in the first case, no threat to health is anticipated so long as the exposure does not induce a disturbance of a kind and degree that overloads the normal protective mechanisms of the body; in the second, a potential for ill health is said to exist as soon as the organism undergoes the first detectable change, of any kind, from its normal state.

A distinctive feature of manganese toxicity is the apparent central role of particle size.[217] The manganese dioxide aerosols encountered occupationally may be classified either as fumes, which consist of particles less than 1 μm in diameter, or as dust, characterized as suspensions of particles whose diameter generally exceeds 5 μm. Chronic exposure to relatively high concentrations of manganese dioxide dust—e.g., as experienced by pneumatic drillers—is associated with extensive impairment of the central nervous system.[444] In contrast, a much smaller exposure to manganese dioxide fume may produce the same symptoms in those who melt or cut the metal. The dependence of deposition sites and solubility rates on particle size and the resulting difference in etiologic role between manganese dioxide dust and fume imply that a unitary threshold limit value is inherently inadequate. The standards should reflect explicitly the key role of particle size.

9

Neurobehavioral Effects of Manganese Deficiency and Toxicity

DEFICIENCY

Although the toxicity of manganese is well known, its presence in the diet is essential. As early as 1939, Norris and Caskey[378] reported congenital ataxia in chicks resulting from a deficiency of manganese in the hen. Ataxia has also been observed in the offspring of manganese-deficient rats,[232,247,451] swine,[407] and guinea pigs.[161] Manganese-deprived mice fail to maintain orientation when submerged in water.[157] Despite the absence of histologic findings or abnormal assays of various enzymes, there are consistent findings of incoordination, lack of equilibrium, and retraction of the head in young animals born of deprived mothers. Hurley and Everson[246] also showed that body-righting reflexes were markedly delayed in the offspring of manganese-deficient rats; this implies that the vestibular apparatus fails to develop properly in their newborn young. More than half such offspring displayed slower development of the equilibrium-related osseous labyrinth than was seen in any of 22 normal newborns.[250] The primary defect is due to an absence of utricular and saccular otoliths of the vestibular portion of the inner ear.[157]

Skulls of manganese-deficient newborn rats were shorter, wider, and higher than those of normal controls, both at birth and thereafter,

Neurobehavioral Effects of Manganese Deficiency and Toxicity   117

thereby distorting the morphology of the brain. The shortening of the skull in the newborn was not due to proportionally decreased growth in the cranial bones. The length of the interparietal and parietal bones in deficient newborns was similar to that in controls, and the length of the frontal bone was less reduced than that of the nasal or that of the skull as a whole. Hurley and colleagues attribute the shortening largely to insufficient growth in the basal portion of the skull. The shortening of the skull, occurring simultaneously with dissimilar growth rates in the cranial bones, results in doming of the frontal portion of the skull.

Manganese deficiency also affected the growth of the brain. In absolute terms, the brain weight was decreased; but, relative to body weight, the brains of the deficient animals were larger than those of the controls. The increase in mass was not a product of edema; rather, it was due to maintenance of brain growth in the presence of diminished body growth (as measured by dry weights). The brain is often spared, compared with other organs, when an animal is placed in a nutritionally inadequate environment. Cerebrospinal-fluid pressures in the two groups were not significantly different.

INTOXICATION

The major problem associated with manganese is intoxication due to excessive absorption. The symptomatic expression of excessive absorption is described in Chapter 7. A number of attempts have been made to quantitate the objective signs and subjective symptoms, but the samples of subjects are so diverse that consistency is often lacking. Mental activity is reported to be slowed, judgment impaired, and memory weakened, but intelligence, as measured by the Wexler adult intelligence scale, was normal in the one patient tested.[433]

Russian authorities, in addition to paying attention to the usual pathologic indicators of toxicity,[422] attach great significance to the "higher nervous activity" of animals exposed to noxious gases and dusts. It is their belief that the cerebral cortex is highly sensitive to external factors in the environment. In general, animals (usually rats) are conditioned to respond to a variety of signals and then exposed to the intoxicants. When a minimal toxic exposure is reached, the animals develop "phasic states"; later, learned patterns of responses break down, and individual reflexes disappear; finally, none of the reflex pattern is left. When the animals are more severely affected, the natural responses to the sight and smell of food (sniffing, looking toward the food, and salivating) disappear. Considerable success has been reported with conditioned-response methods in evaluating lead, mercury, and benzene; concentrations of these substances "considerably below that of the maximum permissible for workshops" produced detectable changes that consisted of "gradually increasing weakness of the stimulatory and inhibitory processes followed by the development of protective reflexes."[422 (p.392)]

Considerably less direct experimental work has been done in applying these methods to manganese. Dokuchaeva and Skvortsova[141] stated that the present Russian standards for ambient environment are based on indirect calculations derived from empirically adopted maximal acceptable concentrations for industrial environments. Because manganese compounds are odorless and have no easily recognized irritating properties, and because numerous problems attend experimentation with airborne dusts, it is understandable why few data are available.

Neurophysiologic and Behavioral Studies in Man

Chronaxies of electrically excitable elements are used as indicators of altered physiologic state. Every receptor, nerve, or muscle can be stimulated into action by an appropriate electric impulse. Once a minimal intensity (the threshold) is reached, the stimulated element responds characteristically, and for each element one can construct a strength-duration curve. By convention, the duration of a stimulus that is exactly twice the threshold intensity (called the "rheobase") that is required to stimulate the element to activity defines the chronaxy. Although it is an arbitrarily selected point on the strength–duration curve, Russian authorities pay significant attention to this measure, for they are convinced that it is sensitive to subtle changes in the physiologic state of the chronically exposed organism. Chronaxies differ for different regions of the body; for example, chronaxies of the extensors are longer than those of the flexors. A profile of chronaxies can be constructed from a complete body survey, and each patient's profile can be compared with the norms developed on healthy populations. With a variety of intoxicants, the flexor chronaxies may exceed those of the extensors. Such a change in chronaxies for antagonistic muscles is interpreted by the Russian investigators as reflecting a weakening of the subordinating influence of the central nervous system.

Chronaxy changes were observed in 91% of Rumanian manganese miners:[531] in 79% of "almost completely healthy" miners, in 92% of those suffering from manganoconiosis, and in 98% of those suffering from chronic manganese intoxication. These percentages include the modification of the chronaximetric value, of the "Bourgignon ratio," or of both. Changes were found in both upper and lower limbs. The changes consist mainly of a decrease in the chronaximetric value in the upper limbs and an increase in the lower limbs. Wassermann and Mihail[531] note that their data "do not permit [establishment of] a correlation between these modifications and the clinical state of the subjects; they only indicate the presence of a subclinical disorder."

Summarizing the neurologic examination made by one patient's physician, Roldan[431] stated:

Exploration of the characteristics of neuromuscular excitability or motor chronaxy shows: (a) elevated chronaxy of peripheral subordination, especially in the muscles of the ectromelic or distal regions of the lower limbs, (b) altered chronaxy of association, by a double mechanism: principally by the alteration of the chronaxy of subordination and possibly also by a central lesion. This indicates, particularly, elevations corresponding to segments L2, L3, and L4 of 0.90 to 2.50 milliseconds (normal: 0.10 to 0.16), increase also in L3, L4, and L5 of 3.0 to 3.22 (normal: 0.22 to 0.36), and increases in L5, S1 and S2 of 1.90 to 3.0 (normal: 0.44 to 0.72).

The physician also noted that systematic exploration of the chronaxy of the superficial cutaneous sensitivity shows zones of discrete hypoesthesia that in general terms are in the ectromelic or distal portions, especially of the lower limbs.

These findings of changes in the patterns of electric sensitivity of the sensorimotor elements, if reliable, might be very useful in early diagnosis. Unfortunately, Abdel Nabi and Kayed[1] could not confirm those results. They attempted to ascertain, through electromyographic (EMG) and conduction-velocity studies, whether peripheral nerves and their muscular supply are affected in chronic manganese poisoning. Using stimulus intensities equal to tissue threshold levels, as used in chronaxy studies, they found no change in conduction velocity in either the ulnar or the peroneal nerves. They also found no changes in the EMG, except in four cases complicated by pellagra and two by spondylosis, in all of which decreased conduction velocity could be expected.

Ryzhkova et al.[435] (cited in Wassermann and Mihail[531]), in a study of 75 patients suffering from manganism incurred in a manganese mill, found modifications of EMG values and changes in the "functional excitability of the analyzers of the cerebral cortex."

Further chronaxy studies were conducted by Oltramare et al.,[384] who used two chronic-manganism patients who had contracted the illness while employed as arc welders. In their duties, they had to weld the inside seams of large-diameter tubes, which exposed them to the fumes and dusts of burning manganese. Despite nearly 30 years of exposure, the two patients had mild cases: their complaints were primarily subjective, with objective measures being mostly negative. Both patients had been away from work for more than a year at the time of testing. Oltramare *et al.* determined strength-duration functions for both rectangular and triangular pulses on 10 muscles. Although the strength–duration curves for some muscles were normal, others were pathologic. Three kinds of alterations were seen: a lengthening of chronaxy, which indicates hypoexcitability; irregularities in the curves for both rectangular and triangular pulses, which, according to Gillert,[197] denotes that the muscle fiber has lost its nerve connection; and a shortening of the chronaxy, which shows neuromotor hyperexcitability.

Positive EMG results further suggesting partial denervation were also reported by these investigators. Instead of showing the electric manifestations of spatial summation of the electric outputs from different muscle-fiber groups, their patients' records exhibited bursts of activity that were independent of the activity of nearby groups. The authors stressed the fact that only through detailed study of the chronaxies, neuromuscular electric excitability curves, and electromyography was it possible to get indisputable alterations signifying injury to the peripheral nervous system. Data from studies of this kind have great potential for detecting early effects of toxicity. This independent corroboration of the sensitivity of the chronaximetric methods argues strongly for their utility.

Abdel Naby *et al.*[2] performed electroencephalographic examinations of 11 patients, studying both resting (alpha) rhythms and barbiturate-induced "fast" (beta) activity. The records of four patients were normal, two records were borderline, and the remaining five records were abnormal. The authors concluded after detailed examination of the clinical histories that there was no direct correlation between the clinical picture and EEG findings. In a parallel investigation, the EEG was within normal limits for all cases studied by Fuenzalida and Mena[183] and by Dogan and Beritic.[136]

The characteristic multiplicity of the patients' signs and symptoms as described in Chapter 7 does not appear synchronously or in a regular sequence. Furthermore, there are marked individual differences in the expression of the disorder. The question must be raised regarding the separation of symptoms that are due directly to the lesions from symptoms that are attributable to the patient's perception of his symptoms and those with other causes. The tendency for the personality and psychomotor changes to precede the more classic neurologic changes and then to disappear when the latter become evident, despite the progressive growth of the central nervous system lesions, suggests a multifactorial basis of the symptoms.

Data gathered from patients after the fact are not easy to interpret; unfortunately, few experimental studies of manganism in which neurobehavioral changes were reported have been conducted to circumvent these difficulties.

Experimental Studies in Animals

As discussed earlier, both deficiency and excess of manganese have profound neurologic and behavioral effects. Psychologic stresses, by the same token, can modify the physiologic capacity to handle manganese. [For example, Starodubova[464] reported a study of manganese balance in adolescents while they were undergoing normal schooling, taking examinations, and on vacation. Girls had a higher turnover and a lower (39%) retention than did boys (77%) during the learning period. Excretion increased during the examinations, when the boys even went into a negative balance. During vacations, the balance was close to zero, with excretion equaling intake.] Experimental studies of intoxication in animals are few, and most of them are concerned with nonneural changes. Wassermann and Mihail,[531] nevertheless, are of the view that the modification of the electric excitability of the neuromuscular apparatus in manganese patients indicates a lesion of the cerebral functions, especially of the cerebral cortex. In the experimentally produced chronic manganese intoxication of the dog, Wassermann and Mihail[531] reported that the first disorder observed consists of the alteration of the "higher nervous activity," which suggests a decrease in the ability to form conditioned reflexes. The first experimental studies outlining the neurobehavioral effects of manganese intoxication in nonhuman primates were conducted by Mella,[346] who used common rhesus monkeys (*Macaca mulatta*) as subjects. Four monkeys were given manganese dichloride every other day for 18 months, the dosage increasing from 5 to 15 or 25 mg/day. "The monkey at first develops movements which are choreic or choreo-athetoid in type, later passing into a state of rigidity accompanied by disturbances of motility; then appear fine tremors of the hands and finally contracture of the hands with the terminal phalanges extended."

Direct exposure of a monkey (*M. mulatta*) to manganese-ore dust containing 50% manganese dioxide was tried by van Bogaert and Dallemagne[514] for 1 h daily for 100 days. Another monkey was given manganese sulfate (about 10–15 mg daily for 300 days) in its food. The animal exposed to the dust developed neurologic symptoms; the other did not. The animal exposed to the dust showed ataxia, wide-based gait, and intention tremor; paralysis of the hind limbs appeared later. It is worth noting that, 6 months after termination of the treatment, chemical analysis detected no manganese in the organs of the animal poisoned through the respiratory tract. Van Bogaert and Dallemagne found cerebellar atrophy, particularly in the Purkinje cells and granule cells but no cytologic damage to the globus pallidus. Damage was, instead, scattered throughout the central nervous system, including the spinal cord.

Pentschew et al.[397] studied the consequences of two intramuscular injections of manganese dioxide in five mature rhesus monkeys. The first symptoms of intoxication were noticed about 9 months after the first injection, when the animal became very excitable, especially when approached or confronted by people. It jumped, hitting its head on the ceiling of the cage. In jumping, it often fell on its heels or buttocks, instead of landing on its feet. In addition, there was general clumsiness; and in moving about, it showed a resemblance to moderate inebriation. No cogwheel phenomenon, tremor, or involuntary movements were observed.

Wassermann and Mihail,[531] after reviewing the experimental literature, concluded that all experimental studies undertaken between 1921 and the date of their writing (1964) "have always succeeded in reproducing in the animal the morphological lesions of the central nervous system by the administration of manganese compounds if the doses and exposure times were sufficient." That is probably generally true, but the tone of the declaration suggests a belief that there are differences in species sensitivity and in the richness of expression of the intoxication. It is also known, for example, that, even among primates, there are wide species differences in both the outward manifestation of intoxication and the histopathologic effects. The generality of the behavioral and neurologic changes in different species is difficult to assess quantitatively, because some investigators have doubtless overlooked behavioral alterations of potential significance. It is also likely that the symptoms are so subtle in some species that they cannot be noticed without special procedures. In addition, the correlation between symptom and pathology is obscure.

In some species, behavioral changes are evident, even though histologic alterations cannot be found. Squirrel monkeys (*Saimiri sciurea*) that received subcutaneous injections of 200 mg of manganese dioxide two or three times at monthly intervals began showing signs of muscular rigidity, flexor posturing of the extremities, or fine rapid tremors of the distal extremities 2 months after the first injection.[374] Two of the monkeys were unable to climb about their cage when prodded, and one exhibited an exaggerated startle reaction, hitting its head on the cage ceiling when approached—a symptom noted also by Pentschew et al.[397] Some monkeys exhibited "obstinate progression" (compulsive walking, even after encountering an obstacle, such as a wall), which reached near somersault proportions. Despite marked reductions in the dopamine and serotonin content of the caudate nucleus (most pronounced in the animals most afflicted), there were no histologic changes in the cerebral cortex, caudate nucleus, thalamus, hypothalamus, subthalamic nucleus, substantia nigra, cerebellum, or other areas of the lower brain stem, whether in neurones, glial elements, or vascular supply. This is further evidence that clinical and biochemical abnormalities may precede the histologic changes, and it suggests that some of the clinical manifestations are primarily biochemical, not anatomic. This investigation also provides evidence, although perhaps weak, in favor of a hypothesis of individual susceptibility, inasmuch as the animals that showed the most effect received only two injections—not three.

Experimentally induced behavioral changes after manganese intoxication have scarcely been reported and dose–response relations are unknown; nevertheless, of all experimental animals on which observations have been made, the mimicry of human symptoms is greatest in chimpanzees. Single injections in rhesus monkeys are without apparent effect, and repeated intramuscular injections of manganese dioxide at 500 mg/kg of body weight produce only mild dystonia and dyskinesia. The effects are considerably more elaborate in the chimpanzee, despite the inadequate and informal testing performed.[398,419] Approximately 3 months after injection at 500 mg/kg in multiple sites of two young chimpanzees (about 4 and 8 years old), gross activity decreased. The animals held unusual positions and postures for up to 30 sec. In general, their deportment was reminiscent of that before injection, but the time course of any action was slowed somewhat, and vigorous activity was absent. A first sign was that the animals had difficulty, after climbing a fence, in negotiating a return to the ground. Ulcers could be found on the palmar surfaces of the fingers as a product of the prolonged fence-climbing. The lower lip was very labile and expressive, and, although it often droops in the healthy chimpanzee, it drooped more than normal in manganese intoxication. Ten days later, activity was slowed further, with difficulty in eating; this evidently marked the beginning of a rapid downward course of illness. Indeed, by this time, the subjects had become significant nursing burdens. Personalities, meanwhile, were apparently unchanged. A mask-like, motionless expression became obvious, periods of somnolence and possibly unconsciousness were exhibited, and locomotion became nearly impossible. Cogwheel rigidity was absent. Athetoid extension and dystonic posturing of the fingers and arms were evident.

Gibbons (*Hylobates lar*) that receive similar injections show scarcely any change other than a tendency to hang on the fence and to walk bipedally slower than they did before injection.

CONCLUSIONS AND IMPLICATIONS

Practically all that is known about the neurobehavioral toxicity of manganese compounds is derived from retrospective studies of human patients suffering its ill effects. Surveys of industrial populations, although valuable, are subject to many complications of selected (rather than random) samples, incomplete assessment of diverse environmental conditions, and limited evaluation of the physiologic status of the subjects. Moreover, the technologic backgrounds and experience for assessing manganese toxicity are limited and have been developed primarily in the clinical setting. Although much is known about the neuropharmacology and psychopharmacology of other drugs, few studies have been conducted with substances that contain manganese. Accordingly, the information at hand is based primarily on clinical experience with a restricted population exposed in past industrial situations.

Experimental studies in animals, which appear to hold the key to full understanding of the pathophysiology of manganism, are extremely few and limited.

Albino rats and mice may display some of the neuropathologic changes associated with human manganism, but they do not exhibit a wide range of behavioral manifestations of the intoxication, and they are of limited use in evaluating subtle changes. Nonhuman primates, particularly monkeys and apes, seem necessary, as one would expect from Cotzias's hypothesis that melanin in the basal ganglia is essentially involved in their functioning, for the amount of melanin in these structures is greatest in monkeys and apes (aside from man).[107] The few studies in which monkeys and chimpanzees were used[346,374,397,514] involved repeated small doses or very large single parenteral doses—different from human exposures.

Rjazanov,[422] in an important paper, has summarized the Russian program of toxicity tests used for a wide variety of potential intoxicants and pollutants. In addition to the usual humoral and pathoanatomic evaluation are those which are neurologic or behavioral in emphasis: learning, discrimination, reflexes, sensory thresholds and intermodal interactions, chronaxy, and avoidance. Tests of these functions are said to be highly sensitive to intoxicating or annoying compounds. Clinical histories of chronically manganic patients are replete with histories of paresthesias, sensory defects, mood shifts, and dyskinesias—all denoting perturbations of the normal activity of the central nervous system. It is unfortunate, then, that very few quantitative tests of the types mentioned were conducted for manganese compounds. The need for data in this area is particularly acute, in that Russia has the most stringent standards of all countries whose standards are available. The Russian authorities doubtless had these considerations in mind when they determined their current restrictive standards.

The finding that chronic manganism, like parkinsonism, is improved after administration of levodopa[108,433] implicates the dopamine system in the substantia nigra, striatum, and pallidum, inasmuch as it is known that the amounts of dopamine and other biogenic amines in the basal ganglia are markedly reduced in parkinsonism.[39,239] (However, it should not be forgotten that the concentrations of acetylcholinesterase and choline acetylase in the striatum in healthy persons are equally high and may also be significant.[345]) In view of the sensory, psychiatric, and motor changes observed in chronic manganism, there is great need for the exploitation of tests of the types advocated by the Russian authorities to assess the lower limits of toxicity.

On the one hand, there is a recognition that existing tests are insensitive to early damage in work areas with high concentrations of manganese, especially those having similarly high concentrations of other irritants. On the other hand, perhaps not all the disturbances found in manganese workers can be attributed to manganese itself. In light of the apparent absence of untoward events in workers exposed to the U.S. limit of 5 mg/m^3 for an 8-h day[10] during the last 10 years,[442] there is little immediate cause for alarm. If additional safety factors become desirable, applicants for positions that involve industrial exposure might be screened for manganese turnover to exclude the more susceptible candidates.[348]

There is little in the literature to enable one to assert that there are sex differences in susceptibility, because the predominance of male patients only reflects the numerical superiority of men in the industry. If it is likely that more women will be industrially exposed to manganese in the future than were in the past, consideration will have to be given to the potential effects on pregnant women and on the fetus. On this subject, the data are especially meager: Ryzhkova *et al.*[435] discovered nine persons (including an unspecified number of women) in a manganese-ore mill who had developed manganism; among them were three whose exposures overlapped a period of pregnancy. The fate of the infants is unknown. Therefore, further study is necessary to clarify this area.

10

Manganese Tricarbonyl Compounds

The industrial hazards associated with the handling of various forms of manganese and manganese ores in mining, steel production, and battery fabrication have been dealt with in preceding chapters. The industrial hazards are circumscribed, and control measures are known. Concerns about public health are related to the aerial discharge of manganese substances that are then inhaled by persons who live in the immediate vicinity and to the occasional infiltration of local water systems with significant amounts of manganese, mostly through location of solid-waste disposal facilities or through contamination of water supplies from sanitary landfills.

Manganese carbonyl compounds hold promise as smoke inhibitors and combustion improvers in fuel oils and as octane improvers in automobile gasolines.

Metal carbonyls are formed when carbon monoxide reacts with so-called transition-series elements at appropriate temperatures and pressures. Carbonyls represent a significant medical hazard: at least one, nickel carbonyl, is carcinogenic, and iron carbonyls, like nickel carbonyl, produce respiratory distress, limb weakness, and tremors.[75] Methylcyclopentadienyl manganese tricarbonyl, $CH_3C_5H_4Mn(CO)_3$ or $C_9H_7O_3Mn$, variously referred to as MMT, CI-2 (for "combustion improver-2"), and AK-33X (for "antiknock 33X"), contains 24.7% (by

Manganese Tricarbonyl Compounds

weight) of manganese. It is currently used as an additive to fuel oil for inhibiting smoke formation and improving combustion and as an antiknock additive to gasoline, usually as a supplement to the common lead antiknock compound, rather than as a replacement. If the gasolines to be marketed in the future for use as automobile fuels are required by law to contain no lead or significantly less lead than they now contain, the consumption of MMT might have considerable possibility for growth. MMT is toxic in itself, and its use as a fuel additive results in the discharge of manganese to the air. The ultimate fate of the discharged manganese is not definitely established. Thus, a potential public-health hazard must be considered, in addition to the industrial hazard.

TOXICITY STUDIES

It is useful to distinguish the hazards of industrial exposure from those of environmental exposure. Handlers of manganese tricarbonyl compounds are exposed to them and their fumes directly, whereas the general population is exposed only to their combustion products, which may be significantly different.

Two studies by Arkhipova et al.[25,26] investigated the problem of exposure. In the first study, laboratory rodents were given cyclopentadienyl manganese tricarbonyl (not the methyl form) at 20 mg/kg or more in a single intragastric feeding. The animals that survived the test for several days appeared sluggish and indifferent to some types of external stimulation. White mice were more resistant than white rats and did not die at doses below 70 mg/kg. The threshold of neuromuscular electrostimulability dropped from an average of 9.6 mA to 5.5 mA, indicating increased sensitivity. Similarly, white rats that received 5 mg/kg daily for 2 months showed slightly lowered neuromuscular thresholds.

The second study investigated sensitivity to the compound when it was inhaled. Rats, guinea pigs, and rabbits were exposed in chambers that contained 0.0007-0.002 mg/liter (average, 0.001 mg/liter) for 4 h/day for 7 months. The rats exhibited no outward manifestations of poisoning throughout the entire exposure period. The authors noted that the threshold of neuromuscular excitability rose 7 months after the experiment began. The data presented, however, also showed that the threshold fell to the preexperimental point in the eighth month, which made interpretation difficult. The authors concluded that the danger from acute poisoning by inhalation was not very great.

Toxicity studies have also been conducted in American laboratories with the methyl form of the compound. The potential hazards of accidental oral intake, percutaneous absorption, and inhalation were investigated. Not unexpectedly, the oral LD_{50} for a single dose is variable, differing among and even within species, for it depends on the sex of the animal, the concentration of the material presented, and the nature of the vehicle carrying the compound. For example, there is some belief that the toxicity is higher when peanut oil or kerosene is the vehicle, rather than water, for the compound is relatively insoluble in water. Studies in both the United States and Russia indicate that the compound is not particularly irritating to the skin or to the eyes. Usually, the industrial vehicle that presents the exposure (gasoline, fuel oil, etc.) is as toxic or as irritating as the manganese compound.

Specifically, E. A. Pfitzer, S. O. Witherup, E. E. Larson, and K. L. Stemmer (unpublished data) analyzed the toxicity of MMT in different laboratory species for all three accidental routes of contact and the intravenous route. The toxicity depends on the species (rats are more susceptible than mice, guinea pigs, rabbits, or dogs). Females are more sensitive than males. The data are not entirely consistent, owing in part to the small number of animals used for each experimental combination. The approximate acutely toxic oral dosages of MMT in the different animals are as follows: rats, 9–176 mg/kg; mice, 350 mg/kg; guinea pigs, 900 mg/kg; rabbits, 95 mg/kg; and dogs, 600+ mg/kg. The acute skin LD_{50} of undiluted MMT is about 1,700 mg/kg (24 h of exposure for male rabbits). A 10% solution in peanut oil in contact with the abdominal skin of rats for 6 h is toxic at 665 mg/kg. Finally, inhalation toxicity is a function of both concentration and duration of exposure. A crude approximation to acute lethality is 500 mg/m^3-h; i.e., if the product of the vapor concentration (in milligrams per cubic meter) and the exposure (in hours) exceeds 500, roughly 50% of the exposed laboratory subjects (except for guinea pigs) will die.

Pfitzer *et al.* also conducted a series of studies of repeated inhalation exposures. Mice, rats, guinea pigs, rabbits, cats, and dogs were exposed for 7 h/day, 5 days/week, up to 30 weeks. Concentrations of 14–17 mg/m^3 produced mortality in rats and mice but not in the other species. Lower concentrations for comparable exposure periods produced no mortality in any species.

They also found that addition of the compound to gasoline at concentrations up to 16 mg/ml and repeated application to the skin produced no adverse effects that were not attributable to the gasoline itself.

Signs of toxic response appeared promptly after exposure by all routes and included mild excitement and hyperactivity, tremors, severe tonic spasms, weakness, slow and labored respiration, occasional mild clonic convulsions, and terminal coma. The animals that survived the convulsive episodes failed to thrive, lost weight rapidly, and died after a few days. With sublethal exposures, after undergoing temporary weight loss, some appeared to recover without sequelae in 2–6 weeks. As expected, the primary pathologic effects occurred in the kidneys and livers of animals that died or were sacrificed after oral, cutaneous, or intravenous exposure. Inhalation at lethal concentrations produced pulmonary changes as well.

As a result of these investigations, a threshold limit value of 0.2 mg/m^3, or 0.1 ppm (expressed as manganese), was proposed in 1970 for industrial exposure.

PUBLIC-HEALTH HAZARD

Smoke Inhibition

In a study of fuel additives used for controlling air pollution from distillate-oil-burning systems, the Office of Air Programs, Environmental Protection Agency, determined the emission of manganese associated with the addition of MMT to a No. 2 oil at a concentration of 1:9,000 to be 1,219 µg/m^3, or 20 mg/kg of fuel.[328] Speaking generally for seven fuel additives tested, the report states that "an analysis of the particulate forms emitted revealed in nearly all cases metals in additives are emitted as metal oxides." Elsewhere, the same report states that several of the metal-containing additives, including MMT, reduced particulate emissions but questions the advisability of their use because the toxicity of the emissions is unknown. A producer of MMT claims that "normal combustion processes convert it readily and completely to innocuous oxides" and reports that "during combustion of these fuels the MMT is converted to solid manganese oxide (our studies usually showing Mn$_3$O$_4$ as product)" [Ethyl Corporation communication to the National Research Council (NRC)]. Manganese trioxide has also been found (Ethyl Corporation communication to the NRC).

Other studies of the use of an unnamed manganese additive in distillate oils fueling gas turbines disclosed small deposits of manganese trioxide and manganese sulfate monohydrate in the hot-gas path on disassembly of the turbine after running 36 h with a fuel manganese content of 100 ppm. "Stack analyses were attempted, but oxides of manganese could not be detected. This is not surprising since the concentration calculated to be about 1 ppm. Analyses for oxides of sulfur and nitrogen were made also, and it appears that there may be a beneficial effect on oxides of nitrogen by use of the additive."[483] The aforementioned EPA study, however, reported that none of the additives reduced nitrogen oxide emissions in the test program.[328]

Ethyl Corporation is apparently the sole producer of MMT, except for experimental quantities made in the USSR. There are other organic manganese compounds on the market as combustion improvers, but the quantities used reportedly amount at most to only one tenth that of MMT. Production of MMT for the last 5 years has been at the rate of 450,000 kg/year, made at a multiproduct facility at Orangeburg, S.C. It is claimed that, as an additive for fuel oils fed to turbines, it provides an environmental benefit in nearly eliminating visible smoke. When it is used as an additive to oils used as boiler fuel, the smoke is less dense, sulfur trioxide and its associated "acid rain" are reduced, and deposits on the boiler tubes are reduced, are more easily removed, and are less corrosive (Ethyl Corporation communication to the NRC). A maximal manganese emission of 0.3 kg/h has been estimated for a power plant using MMT (Ethyl Corporation communication to the NRC).

Gasoline Additives

If manganese compounds do indeed become commonplace for their antiknock capability, the amount of their combustion products in the atmosphere will increase considerably. Toxicity will depend in part on the precise nature of the combustion products, for more than 99% of the organic manganese is consumed in combustion, and the exhaust discharge is therefore inorganic (Ethyl Corporation communication to the NRC). No long-term studies for carcinogenicity of the exhaust gases are available. What the exhaust products will be is not known exactly, but manganese tetroxide, believed to be one of the more toxic manganese oxides, will be one of them—doubtless the predominant one. The form of the manganese and the possibility of chemical reaction that results in the formation of hazardous products in the atmosphere are disputed. The absence of definitive quantitative data on their toxicity points to the necessity for further research, inasmuch as such data will be needed before the use of these substances becomes widespread.

If MMT were used as an antiknock supplement in 50% of the gasoline used in the United States in the proportion of ¼ g of manganese per gallon of gasoline, the producer estimates that "the most this would add to urban air would be 0.05 to 0.2 micrograms of Mn/m^3" (Ethyl Corporation communication to the NRC). This is based on two studies [319,485] of lead in urban air and the assumption that manganese and lead emissions to the atmosphere are proportionate, with the manganese being one twentieth the lead. If MMT were to be used as a primary antiknock—

that is, in the range of 1.0-2.0 g of manganese per gallon—then, by extrapolation, emission of manganese to the atmosphere would be increased to ranges of 0.2-0.8 to 0.4-1.6 $\mu g/m^3$. However, problems of spark-plug life and exhaust-valve life that accompany use at high manganese concentrations suggest that it is unlikely that a concentration of more than ½ g/gal will be used.

CONCLUSIONS

As is the case for other forms of manganese exposure, more is known about industrial hazards of MMT than about its public hazards. The exhaust products of MMT combustion are specific forms of manganese. Experience with it is extremely limited. No information is available on very long exposures at low concentrations or on exposures of special groups, such as the aged, the chronically ill, the pregnant, or those under chronic medication. The general subject of the relation of the organometallic manganese fuel additives to the problems of air pollution appears to require considerable further study.

11

Conclusions

MANGANESE IN THE ECOSYSTEM

Manganese is one of the more abundant mentallic elements in the earth's crust. It occurs in constantly changing complex relations involving physical, chemical, and biologic activity, with the oxides being the dominant chemical forms. It is present in nearly all, if not all, organisms.

Manganese is used primarily in the production of iron and steel, and the emission of manganese to the atmosphere is principally the result, either direct or indirect, of this use. However, reliable recent data are often lacking. Available data must be used with caution and must take into account various qualifying factors, including location, atmospheric conditions, degree of control, and method of sampling and analysis. Maximal manganese concentrations in ambient air in the United States, occurring at industrial urban locations, have exceeded 10 $\mu g/m^3$ but appear to have been of that general magnitude. The average urban ambient-air concentration in the United States is approximately 0.10 $\mu g/m^3$ for a 12-year period ending in 1965. Widespread use of manganese organometallic compounds as gasoline additives and as combustion improvers and smoke depressants for fuel oils might present a pollution problem in the future. High concentrations of manganese in waters, resulting from mining and industrial pollution, are soon dissipated in the

Conclusions

rivers, and there are great differences in manganese concentration between samples of water taken at the same locations but at different times. There are few reports of man-made manganese contamination of soils.

Conclusive data are lacking too often for a definitive evaluation of manganese in the ecosystem. However, except for the evaluation of manganese organometallic compounds as possible future air pollutants, currently available data appear to be adequate for the general purposes of this report. The following conclusions seem reasonable:

1. High concentrations of manganese in the atmosphere are usually the result of man's activities.
2. Current manganese pollution of the atmosphere is a local problem—for the most part an in-plant, rather than an ambient-air, problem.
3. Widespread use of manganese fuel additives might create a more general ambient-air problem. This possibility requires objective evaluation before the additives come into general use.
4. Manganese pollution of water appears not to be a problem, except under very unusual local circumstances.

MANGANESE IN PLANTS

The availability of manganese to a given crop is determined by many soil factors. These include concentrations of total, soluble, and easily reducible manganese; pH; concentrations of other cations, anions, and total salts; cation-exchange capacity; organic-matter content; drainage; compaction; temperature; and microbial activity. The utilization of manganese absorbed by a given plant is affected by various factors, including temperature, light intensity, and interactions between manganese and iron or other elements within the plant. These soil and plant factors determine whether manganese concentrations will be deficient, adequate, or excessive for the growth of a given plant. Furthermore, plant species and varieties differ widely in manganese requirements, accumulating ability, and tolerance to excessive manganese within their tissues.

Manganese contents and yields of crops can be regulated by soil management practices that alter the concentrations of divalent manganese in soils, by the application of manganese-containing foliar sprays, and by choice of crop species or variety. Evidence that manganese uptake and use are genetically controlled suggests that varieties can be bred for greater tolerance to low- or high-manganese environments that cannot be modified economically. Such varieties would be useful in maintaining crop yields and in regulating the manganese contents of plant products.

Thanks to recent developments in analytic techniques, especially atomic-absorption spectrometry and neutron-activation analysis, methods for determining manganese in all aspects of man's environment are available or capable of readily being made available. The one serious deficiency is the lack of a good bioassay method for evaluating the physiologic effects of relatively small increases or decreases in the exposure of men and other organisms to manganese in the environment.

MANGANESE IN MAN AND OTHER ANIMALS

Manganese is an essential element for man and other animals. Unequivocal evidence of deficiency in man has not been presented. Most human food contains measurable amounts of manganese; the richest sources are nuts, cereal products, dried legume seeds, green leafy vegetables, and dried fruits. Animal tissues and dairy and fish products are very low in manganese. The percentage of ingested manganese that is absorbed apparently is low.

The major route of absorption—i.e., whether respiratory or gastrointestinal—has not been established, although many feel that the respiratory route is the most important.

Excretion takes place mainly through the intestinal wall and via the bile.

The body of a normal 70-kg person contains about 15–20 mg, widely distributed but more concentrated in the pituitary, pancreas, liver, kidneys, and pigmented tissues, including hair.

The biochemistry of manganese has not been determined completely; however, manganese probably has multiple biochemical roles. It influences many enzyme activities *in vitro*; some enzymes, such as succinic dehydrogenase, demand manganese. The mitochondrial enzyme pyruvate decarboxylase, which catalyzes the carboxylation of oxaloacetate from pyruvate, is a manganese metalloenzyme. Manganese is an essential micronutrient for all species of birds and mammals that have been investigated. The symptoms of manganese deficiency include ataxia, attributable to faulty embryonic development of the inner ear; severe skeletal deformities; impaired glucose tolerance; and impaired reproductive function. A biochemical basis for the skeletal and connective-tissue defects has been provided by the recent definition of manganese-dependent enzyme systems that are necessary for mucopolysaccharide synthe-

Conclusions

sis. The possible occurrence and sequelae of human manganese deficiency have not been investigated.

Data on the biochemical abnormalities that result from chronic manganese toxicity and are responsible for the clinical manifestations of manganism are very limited. Animal studies, the similarities between manganism and Parkinson's syndrome, and the beneficial effects of L-dopa therapy indicate that a major abnormality may be a defect in the synthesis and storage of dopamine. In some cases, a deficiency of brain serotonin may be more important. Both substances probably are essential for normal function. In Parkinson's syndrome, there is a severe reduction in the melanin content of the substantia nigra. Although it is not certain whether this is also characteristic of chronic manganese toxicity, manganese is concentrated in melanin; and the interrelations of parkinsonism, melanin granules, and manganese may be more than coincidental.

TOXICOLOGY OF MANGANESE

It was not until 1837 that manganese was implicated as a disease producer. After an initial report by Couper, physicians from many countries began reporting a similar disease. The disease occurred in manganese miners, steel-plant workers, and others who had the opportunity to inhale manganese dust or fumes. A form of lobar pneumonia was reported in inhabitants of a Norwegian town contaminated by the effluent of a metallurgic plant producing ferromanganese. This was followed by reports of a similar disease in other countries, although it has not been reported in the United States. Manganic pneumonia is unresponsive to antibiotics.

Intoxication in man is primarily the result of inhalation, although absorption via the gastrointestinal tract and even through the skin has been reported. The most notable effects of excessive exposure result from actions on the central nervous system.

The symptomatic expression of excessive manganese absorption in patients exposed industrially is usually insidious, mimicking a number of other diseases. Psychologic disturbances and disorders of motor coordination are prominent, particularly in the later stages of the disease. General adaptiveness is diminished, but tested intelligence appears to remain unchanged. Russian investigators have presented electrophysiologic data that they interpret as reflecting a weakening of the subordinating influences of the central nervous system and as denoting a subclinical disorder. Electroencephalographic studies, however, have failed to reveal changes attributable to manganese intoxication.

Experimental studies of animal intoxication are few and not behaviorally oriented. Nevertheless, gross behavioral changes are observed in intoxicated animals, chiefly in primates, sometimes in the absence of histologic changes in nervous tissue. Choreic or choreoathetoid movements, rigidity, ataxia, and fine tremor have been reported in both New World and Old World monkeys. Similar, but even more profound, changes have been noted in chimpanzees.

CURRENT STANDARDS

The threshold limit value (TLV) of 5 mg/m^3 recommended by the American Conference of Governmental Industrial Hygienists is generally thought to have a low factor of safety for susceptible persons. It is very difficult to judge the validity of the current TLV from case studies, because there is frequent reference to "the possibility that higher exposures may have occurred." The duration and degree of exposure appear to be important factors in the development of disease in susceptible persons.

Whether a particular manganese compound is implicated in the production of disease is not clear. Some observations appear to indicate that fumes are more likely to produce disease than equivalent quantities of dust.

Individual susceptibility plays a major role in the development of manganism. No substantial clues have been published as to why one person develops the disease while fellow workers in the same environment do not.

Present concentrations of manganese in ambient air, as reported by the National Air Surveillance Networks, appear to provide a substantial factor of safety to the population, except to groups that are close to point sources of large emissions.

The long-term toxicology of manganese, including fetal effects, still presents a collection of ambiguous answers. For this reason, special care must be exercised before substantial additional sources of manganese are introduced into the environment.

12

Recommendations for Future Research

In preparing this report, the Panel discovered that information on many biologic aspects of manganese is either lacking or incomplete. Additional knowledge is needed to assess fully the biologic implications of manganese. The prime requirement is the determination of a safe degree of exposure and a minimal requirement for manganese in the total environment.

The following is a list of subjects that the Panel believes should be given priority in the search for information on manganese. They are not arranged in any order of importance.

1. Is there individual human susceptibility to excessive or deficient concentrations of manganese? If so, how can it be detected, and how can it be predicted? Are the differences due to diet, genetic makeup, concomitant stress, variations in absorption, disease, or interactions with drugs and chemicals? Are there also group differences?

2. What are the effects on pregnant women and infants of chronic excessive exposure to manganese? Is the fetus at risk?

3. What controls the metabolism and turnover of manganese?

4. Is there true reversibility at any stage in manganism? Does one exposure increase susceptibility to a second exposure?

5. What accounts for the time course of the symptoms in manganism? Why do the psychiatric symptoms precede the neurologic?

6. With few exceptions, manganese pollution does not occur in isolation from pollution from other substances. How do these pollutants interact? Are their effects merely additive, or do some combinations create special hazards to health?

7. How valid are the neurobehavioral tests used by Russian investigators? Are they so generally responsive to external stimuli that they only indicate labile normal functioning, or are they diagnostic of truly pathologic states?

8. There is an obvious need for evaluation of mood and memory changes in manganese intoxication. The depletion of dopamine and perhaps other substances by manganese or endogenous substances produces effects that extend far beyond the basal ganglia, as indicated by changes in mood and memory.

9. Are the so-called lower oxidative states more toxic than the higher ones? This has often been reported but has not been proved. Indeed, dose–response relations have not been established for any manganese compound. Does the toxicity of manganese depend on its physical form?

10. There is a need for additional epidemiologic investigation of manganism and respiratory diseases as related to manganese exposure.

11. Animal and human studies are needed on the effects of various chelating agents—calcium-disodium-EDTA, penicillamine, and others—on early and late cases of manganese poisoning and on healthy controls.

12. There is a need for reexamination of the relative importance of the lungs, as opposed to the gastrointestinal tract, as a route of absorption of manganese from the atmosphere.

13. Further research is needed to determine the clinical value of present tests of blood, urine, and hair as indices of recent absorption of excessive manganese. Does increased manganese content of any of these samples correlate with later features of manganese toxicity? Is manganese in hair entirely endogenous in origin (in which case it may provide a valid index of manganese absorption and body stores) or can direct atmospheric contamination of the exposed hair shaft contribute to the analytic results? Is there correlation between these indices and the concentration, particle size, and chemical state of the manganese in the ambient air?

14. There is a need for research on the human requirement for manganese and the incidence of nutritional manganese deficiency. The range of requirements for manganese is incomplete, and the incidence of nutritional manganese deficiency in humans has never been investigated. There is evidence that genetic and metabolic abnormalities may result in increased nutritional demand for manganese. Additional research should therefore be conducted to establish the range of nutritional re-

Recommendations for Future Research

quirements for manganese, to perfect means for detecting human manganese deficiency, and to determine the incidence of manganese deficiency.

15. The chemical and physical nature, quantity, and toxicity of the emissions resulting from the various uses of methylcyclopentadienyl manganese carbonyl and other manganese organometallic substances used as fuel additives should be determined, as well as the nature, quantity, and toxicity of reaction products that could result from the emission of such substances to the atmosphere.

16. Studies on workers in contact with manganese carbonyl compounds during production and evaluation of such studies that have already been done are necessary.

17. The relation between manganese toxicity and manganese-induced iron deficiency in plants should be clarified.

18. What are the possible roles of high internal phosphorus and calcium content in the detoxification of manganese within plant tissues?

19. By what physiologic mechanisms may molybdenum reduce or accentuate manganese injury in plants?

20. By what mechanisms do light intensity and temperature affect the toxicity of a given internal manganese concentration in plants?

21. What is the genetic basis of differential manganese requirements and differential tolerance of excess manganese among plant species and varieties? What is the genetic basis of the later development of plant varieties adapted to low or high concentrations of available manganese in the growth medium?

22. What are the relations of differential manganese requirements and differential tolerance of excess manganese to wet-soil tolerance, pH and other properties of root excretions, oxidizing and reducing capacities of roots, auxin and enzyme activities, amino and organic acid balance, and physical or chemical compartmentalization of manganese within plants?

23. Does any plant material ever contain sufficient manganese to injure animals or man?

24. Do plants absorb manganese from the air?

25. What are the manganese requirements and tolerances of soil microorganisms?

Appendix: Environmental Sampling and Analysis of Manganese

Investigation into the biologic effects of manganese as an atmospheric pollutant requires methods for sampling and analyzing traces of manganese in air, water, soil, foods, vegetation, biologic fluids and tissues, and a wide variety of other materials. Fortunately, as a result of recent advances in analytic chemistry, there are methods at hand that are adequate for almost any problem to be encountered, with the possible exception of the determination of volatile organic manganese compounds in the atmosphere. Although analytic methods for air, biologic samples, and other materials often overlap, for the sake of convenience the various areas of interest are divided here rather arbitrarily into air sampling and analysis, analysis of biologic samples, and analysis of all other types of materials.

INORGANIC MANGANESE PARTICLES IN AIR

In-Plant Sampling

DUSTS AND MISTS

For relatively coarse particles, such as the usual industrial dusts and mists, with a particle size of approximately 1–10 μm, the midget impinger, the standard impinger, and the electrostatic precipitator may be used for the collection of suitable samples.[11]

FUMES

In general, impingers are not suited for the efficient collection of finely divided manganese-containing aerosols, such as the fumes from welding, oxyacetylene cutting operations, and the pouring of molten metal. For this type of sample, glass-fiber filters with a collection efficiency of at least 99% for particles 0.3 µm in diameter[343] or organic-membrane filters of equal efficiency are recommended.[252,487] The electrostatic precipitator can also be used, but it is less convenient.[262]

AMBIENT AIR

The sampling method most commonly used for ambient air is the 20 × 25-cm glass-fiber filter with a high-volume air sampler.[343] However, the glass-fiber filters have the drawback that the blanks for trace metals are sometimes very high and may be rather variable from filter to filter. Also, quantitative recovery of trace metals from the glass-fiber filter may be difficult.

Recently, it has been demonstrated that membrane filters are excellent for sampling ambient air for most trace metals, including manganese, although the air flow for a given face area cannot be as high as for the glass-fiber filter and the total quantity of particles collected is therefore somewhat less for a given sampling time. The necessity for using a reduced flow rate is more than offset by the low blanks for trace metals in the organic-membrane filters and the ease of recovery of minute quantities of metals.[252,487]

Analysis

SAMPLE PREPARATION

The preparation of impinger and electrostatic-precipitator samples for analysis is discussed elsewhere.[11] The treatment of glass-fiber filters is described in some detail by Thompson et al.[487] These investigators have pointed out the advantages of the low-temperature oxygen asher, compared with the conventional muffle furnace, with respect to loss of the trace elements while the samples are being ashed.

PERIODATE METHOD

The periodate method is the classic wet chemical method of analyzing air samples for manganese.[11,484] It has the advantage that it can be used

Appendix: Environmental Sampling and Analysis 143

in almost any chemical laboratory with relatively simple equipment. The final colorimetric estimation can be made satisfactorily with Nessler tubes, if necessary. However, it requires considerable skill and attention to detail to get satisfactory results. It is difficult to get and maintain complete oxidation of the manganese to permanganate at very low concentrations. Also, the sensitivity is rather poor, in comparison with that of other methods.

SPECTROGRAPHIC METHOD

The spectrographic method for the determination of trace metals, including manganese, in air samples and a wide variety of other materials has long been known.[88,95,476,492] Carlberg et al.[88] and Tipton et al.[492] describe internal standard methods that, with suitable variations in sample preparation, can be used equally well for air or biologic samples in the microgram range. Cholak and Hubbard[95] describe a spectrochemical method in which the manganese is isolated from interferences and concentrated in a small volume by complexing with sodium diethyldithiocarbamate and extracting with chloroform before analyzing with the spectrograph. Tabor and Warren[476] discuss briefly a semiquantitative method suitable for the estimation of trace metals, including manganese, in samples collected on large glass-fiber filters such as are commonly used in community-air studies.

The spectrographic method has been and is still being used extensively. It has the advantages that it can be made specific or nearly so for almost any element, has adequate sensitivity for most types of air samples, and can be used to determine a number of elements concurrently on the same sample. A disadvantage is that a substantial investment in space and money is necessary. Also, a high degree of skill is needed in this rather specialized field of analysis.

NEUTRON-ACTIVATION ANALYSIS

Neutron-activation analysis, a relatively new technique, has been found most suitable for the analysis of very low concentrations (nanogram range) of manganese. The use of this method for air samples is described by Nifong et al.[377] and Dams et al.[124]

Neutron-activation techniques are generally more sensitive than others. This is certainly the case where trace concentrations of manganese are concerned. The principal disadvantage is that it is necessary to have access to a suitable neutron source.

ATOMIC-ABSORPTION ANALYSIS

The use of atomic-absorption analysis for trace metals, including manganese, in the atmosphere has been described recently.[252,487] The atomic-absorption method has a number of advantages over the older methods. It is relatively simple to use and is highly specific for a given element. Its sensitivity is at least as good as, and in many cases better than, that of other methods, except neutron activation. Moreover, it is fairly free of interferences, except for possible matrix effects, which can generally be avoided by dilution of the sample solution so that the dissolved-solids content is less than 0.5%. When glass-fiber filters are used, silica extracted from the fibers can cause interferences with the determination of manganese, zinc, iron, and possibly other elements, unless removed before the sample is subjected to atomic-absorption analysis.[487]

ORGANIC MANGANESE COMPOUNDS IN AIR

Although it is rather unlikely that manganese could be present in the organic form in the ambient air, it is desirable that a method of air sampling and analysis for these compounds be available. Unfortunately, there does not appear to be any such method at hand in the scientific literature. However, it seems logical that some of the procedures now being used for the determination of methylcyclopentadienyl manganese carbonyl in gasoline or other liquid fuels might be modified to be suitable for the determination of these compounds in the atmosphere.[44,80] A number of methods developed for other metal carbonyls could be modified for the determination of manganese carbonyls.[337]

A fairly recent Russian publication[25] mentions a method for determining cyclopentadienyl manganese tricarbonyl vapor in air developed by M. S. Bykhovskaya but gives no reference to the literature or procedural details.

MANGANESE IN BIOLOGIC MATERIALS

Sampling

URINE

The sampling and analysis of urine for manganese have proved rather disappointing as a means of evaluating occupational exposure to manganese aerosols. Persons with signs and symptoms suggestive of manganese

Appendix: Environmental Sampling and Analysis 145

poisoning do not necessarily have high urinary manganese concentrations, and persons exposed to high concentrations do not necessarily have symptoms or high urinary manganese concentrations. However, the group average of urinary concentration in exposed workers exhibits a rough correlation with the average air concentrations.[479] The usual range of urinary manganese concentration for normal adults with no occupational exposure is 1–8 µg/liter.[95,479] Analysis of urine is of value in cases of suspected manganism in which the subjects are to be treated with EDTA. EDTA has the property of mobilizing manganese from the body store and causing it to be excreted in the urine at an increased rate. This technique is useful in distinguishing between parkinsonism of non-occupational origin and manganism due to occupational exposure to manganese aerosols.

The collection of urine samples is described in some detail in an American Public Health Association booklet;[13] although this booklet is intended to apply to lead analysis, the part dealing with sampling would also apply to urinary manganese. For additional information, see *Laboratory Procedures.*[58]

BLOOD

The use of blood as a means of evaluating occupational exposure to manganese has been even more disappointing. There appears to be no relation whatsoever between blood manganese concentration and degree of exposure;[479] an explanation may be provided by the homeostatic mechanism for manganese discussed by Schroeder *et al.*[443]

Blood manganese content for normal adults with no occupational exposure is 2–10 µg/100 g of blood (mean, 4.0–5.0 µg/100 g).[95]

The collection of blood samples for manganese analysis is described in *Laboratory Procedures.*[58]

OTHER TISSUES

Methods of collecting and handling samples of other tissues have been described by Tipton *et al.*[492] Carlberg *et al.*[88] have discussed handling of lung tissue.

Analysis

SAMPLE PREPARATION

With the exception of neutron-activation analysis, the preparation of all biologic materials requires destruction of the organic matter by wet or dry ashing or by the use of the low-temperature asher.[487] In the case of manganese, there is little danger of loss of the element by either dry or wet ashing, so the low-temperature asher is not necessary, although it may be more convenient than the older methods.

The ashed samples are usually taken up in a few milliliters of dilute acid. The preparation of samples has been described in some detail by Carlberg et al.,[88] Cholak and Hubbard,[95] Butt et al.,[83] and Tipton et al.[492]

PERIODATE METHOD

The periodate method has been used for the analysis of biologic samples, but it is definitely not the method of choice, primarily because of its poor sensitivity.[95,106]

SPECTROGRAPHIC METHOD

A number of investigators have used a direct spectrographic technique that involves the concentration of the sample to a very small volume and the use of an internal standard.[88,492] However, the spectrographic method, when used in this manner, does not have quite the sensitivity desired for such analysis. Cholak and Hubbard were able to improve the sensitivity of the spectrographic method considerably by first removing the iron (in blood, liver, etc.) with Cupferron (a chelating agent), complexing the manganese with diethyldithiocarbamate, and then extracting with chloroform and concentrating to a small volume in nitric acid.[95]

NEUTRON-ACTIVATION ANALYSIS

Neutron-activation analysis is by far the most sensitive method for manganese and has been successfully used for analysis of a number of biologic materials. This method is unique, in that the sample is first subjected to neutron bombardment in a nuclear reactor. Then the radioactive manganese[262,487] so formed is separated from other radioactive elements by flash ashing with oxygen, dissolving in acid, oxidizing to permanganate, precipitating with tetraphenylarsonium chloride, and measuring the radioactivity.[105 (pp. 406-407)] Although the neutron-activation technique is both sensitive and specific, as used by Cotzias, it suffers from the handicap that suitable neutron sources are not readily available.

Appendix: Environmental Sampling and Analysis 147

ATOMIC-ABSORPTION ANALYSIS

Atomic-absorption analysis of biologic samples, although not quite as sensitive as neutron activation, is more readily accessible. When combined with some preliminary concentration and removal of potential interferences, it has proved satisfactory for a number of investigators.[7,524] Mahoney et al.[324] have described an atomic-absorption method for serum manganese that, in the absence of high concentrations of iron, does not require preliminary chemical treatment of the sample.

BIOASSAY TECHNIQUES

One serious deficiency that is hampering research into the biologic effects of manganese is the lack of a good bioassay technique. Because the analysis of blood and urine samples for manganese has proved so disappointing, techniques to accomplish this end must be sought. At least three possibilities seem worthy of investigation: sampling and analysis of human hair for manganese, determination of cystathionase activity in blood, and the analysis of blood or urine for catecholamines or for enzymes essential to the metabolism of catecholamines.

Manganese Content of Hair There is some reason to believe that manganese may concentrate in the hair somewhat as arsenic does as a result of exposure to increased concentrations.[47,433] Apparently, the manganese concentration may be higher in chest hair than in scalp hair.[433] There is a complicating factor: the manganese in hair and other keratinous structures seems to be concentrated in the melanin-containing pigmented granules, with very low manganese concentrations found in white hair, fingernails, and other nonpigmented structures.[114] Nevertheless, the possibility that manganese hair concentrations are related to degree of exposure seems worthy of investigation. The collection and preparation of hair samples have been described in connection with the neutron-activation method of analysis.[114] Petering et al.[401] have discussed sampling and analysis of hair for trace metals by an atomic-absorption method.

Cystathionase Activity Another technique for bioassay has been proposed and described by Khalatcheva and Boyadjiev.[279] These investigators have reported that the activity of cystathionase in rats is considerably decreased after poisoning with injection of manganese as manganese dichloride. The phenomenon is said to be reversible, with cystathionase activity returning to normal 60 days after the termination of manganese administration.

Urinary Excretion of Catecholamines There is some indication that the determination of dopamine and other catecholamines may be useful in the evaluation of exposure to manganese. Barbeau et al.[40] have stated that the urinary excretion of dopamine was significantly lower in a group of 16 patients with parkinsonism than in normal controls. The close parallelism of manganism and parkinsonism have often been noted. Recently, Papavasiliou and co-workers have developed strong indications that the metabolism of manganese is linked with that of the biogenic amines by means of cyclic 3',5'-adenosine monophosphate.[390] Thus, it is conceivable that the excessive absorption of manganese from the environment, whether of occupational origin or not, would be reflected in some subtle change in the concentration of one or more of the catecholamines or in the activity of some enzyme needed for catecholamine metabolism. Methods are available for the sampling and analysis of blood and urine for a number of the catecholamines and their metabolites.[304]

MANGANESE IN MISCELLANEOUS MATERIALS

The sampling and analysis of water for traces of manganese and other materials are fairly well covered by Taras et al.[480] Additional information may be found in Foltz et al.,[173] Hines and Dulski,[233] and Fishman and Erdmann.[168] The neutron-activation analysis of water samples has been discussed in considerable detail by Bhagat et al.[59]

Recent methods of analysis of various foods for trace metals, including manganese, have been reviewed elsewhere.[168,173,233] And recent developments in the analysis of alloys, slags, and ores by atomic absorption and neutron activation were discussed by the same authors.[168,173,233]

References

1. Abdel Nabi, S., and K. S. Kayed. EMG and conduction velocity studies in chronic manganese poisoning. Acta Neurol. Scand. 41:159–162, 1965.
2. Abdel Naby, S., K. S. Kayed, and M. A. Aref. EEG induced fast activity in chronic manganese poisoning. Acta Neurol. Scand. 40:259–268, 1964.
3. Abernethey, R. F., M. J. Peterson, and F. H. Gibson. Spectrochemical Analyses of Coal Ash for Trace Elements. Bureau of Mines Report of Investigations 7281. Pittsburgh, Pa.: U.S. Department of Interior, 1969. 30 pp.
4. Adams, F. Manganese, pp. 1011–1018. In C. A. Black, Editor-in-Chief, and D. D. Evans, J. L. White, L. E. Ensminger, and F. E. Clark, Associate Eds. Methods of Soil Analysis. Part 2. Chemical and Microbiological Properties. Number 9 in the series Agronomy. Madison, Wisc.: American Society of Agronomy, Inc., 1965.
5. Adams, F., and R. W. Pearson. Crop response to lime in the Southern United States and Puerto Rico, pp. 161–206. In R. W. Pearson and F. Adams, Eds., R. C. Dinauer, Managing Ed. Soil Acidity and Liming. Number 12 in the series Agronomy. Madison, Wisc.: American Society of Agronomy, 1967.
6. Adams, F., and J. I. Wear. Manganese toxicity and soil acidity in relation to crinkle leaf of cotton. Soil Sci. Soc. Amer. Proc. 21:305–308, 1957.
7. Ajemian, R. S., and N. E. Whitman. Determination of manganese in urine by atomic absorption spectrometry. Amer. Ind. Hyg. Assoc. J. 30:52–56, 1969.
8. Amdur, B. H., H. Rilling, and K. Bloch. The enzymic conversion of mevalonic acid to squalene. J. Amer. Chem. Soc. 79:2646–2647, 1957.
9. Amdur, M. O., L. C. Norris, and G. F. Heuser. The lipotropic action of manganese. J. Biol. Chem. 164:783–784, 1946.

10. American Conference of Governmental Industrial Hygienists. C-manganese and compounds (as Mn), pp. 149–150. In Documentation of the Threshold Limit Values for Substances in Workroom Air (3rd ed.). Cincinnati: American Conference of Governmental Industrial Hygienists, 1971.
11. American Conference of Governmental Industrial Hygienists, Committee on Recommended Analytical Methods. Determination of manganese in air, periodate oxidation method, pp. Mn-1–Mn-4. In Manual of Analytical Methods Recommended for Sampling and Analysis of Atmospheric Contaminants. Cincinnati: American Conference of Governmental Industrial Hygienists, 1958.
12. American Iron and Steel Institute. Annual Statistical Report, 1970. Washington, D.C.: American Iron and Steel Institute, 1971. 80 pp.
13. American Public Health Association, Inc., Occupational Health Section. Methods for Determining Lead in Air and in Biological Materials. New York: American Public Health Association, Inc., 1955. 69 pp.
14. Anderson, I., and H. J. Evans. Effect of manganese and certain other metal cations on isocitric dehydrogenase and malic enzyme activities in Phaseolus vulgaris. Plant Physiol. 31:22–28, 1956.
15. Anderson, J. M., N. K. Boardman, and D. J. David. Trace metal composition of fractions obtained by digitonin fragmentation of spinach chloroplasts. Biochem. Biophys. Res. Commun. 17:685–689, 1964.
16. Anderson, M. S. Sewage Sludge for Soil Improvement. U.S. Department of Agriculture Circular 972. Washington, D.C.: U.S. Government Printing Office, 1955. 27 pp.
17. Anderson, O. E., and F. C. Boswell. Boron and manganese effects on cotton yield, lint quality and earliness of harvest. Agron. J. 60:488–493, 1968.
18. Anderson, O. E., and R. M. Harrison. Micronutrient variation within cotton leaf tissue as related to variety and soil location. Soil Sci. Plant Anal. 1:163–172, 1970.
19. Andrew, C. S., and M. P. Hegarty. Comparative responses to manganese excess of eight tropical and four temperate pasture legume species. Aust. J. Agric. Res. 20:687–696, 1969.
20. Anke, M., and B. Groppel. Manganese deficiency and radioisotope studies on manganese metabolism, pp. 133–136. In C. F. Mills, Ed. Trace Element Metabolism in Animals. Proceedings of WAAP/IBP International Symposium, Aberdeen, Scotland, July 1969. Edinburgh: E. & S. Livingstone, 1970.
21. Ansola, J., E. Uiberall, and E. Escudero. La intoxicacion por manganeso en Chile (Estudio sobre 64 casos). 1-a communicación: Factores ambientales y etiopatogenia. Rev. Med. Chile 72:222–229, 1944.
22. Ansola, J., E. Uiberall, and E. Escudero. La intoxicacion por manganeso en Chile (Estudio sobre 64 casos). 2-a communicación: Aspectos clinicos, incapacidad y reparación médico-legal. Rev. Med. Chile 72:311–322, 1944.
23. Apgar, J. Effect of zinc deficiency on parturition in the rat. Amer. J. Physiol. 215:160–163, 1968.
24. Arber, W., and S. Linn. DNA modification and restriction. Ann. Rev. Biochem. 38:467–500, 1969.
25. Arkhipova, O. G., M. S. Tolgskaya, and T. A. Kochetkova. Toxicity within a factory of the vapor of new antiknock compound, manganese cyclopentadienyltricarbonyl. Hyg. Sanit. 30:40–44, Apr.–Jun., 1965.

References

26. Arkhipova, O. G., M. S. Tolgskaya, and T. A. Kochetkova. Toxic properties of manganese cyclopentadienyltricarbonyl antiknock substance. U.S.S.R. Literature on Air Pollution and Related Diseases 12:85–89, 1966.
27. Asling, C. W., and L. S. Hurley. The influence of trace elements on the skeleton. Clin. Orthop. Rel. Res. 27:213–264, 1963.
28. Asling, C. W., L. S. Hurley, and E. Wooten. Abnormal development of the otic labyrinth in young rats following maternal dietary manganese deficiency. VII. International Congress of Anatomists and 73rd Annual Meeting, American Association of Anatomists, April 11–16, 1960. Abstracts of papers from platform. Anat. Rec. 136:157, 1960.
29. Atkinson, R. L., J. W. Bradley, J. R. Couch, and J. H. Quisenberry. Effect of various levels of manganese on the reproductive performance of turkeys. Poult. Sci. 46:472–475, 1967.
30. Axelrod, J. O-Methylation of epinephrine and other catechols in vitro and in vivo. Science 126:400–401, 1957.
31. Axelrod, J., W. Albers, and C. D. Clemente. Distribution of catechol-O-methyl transferase in the nervous system and other tissues. J. Neurochem. 5:68–72, 1959.
32. Axelrod, J., and R. Tomchick. Enzymatic o-methylation of epinephrine and other catechols. J. Biol. Chem. 233:702–705, 1958.
33. Baader, E. W. Eine Reistudie über schweren Manganismus bei ägyptischen Mangangrubenarbeitern. Arch. Gewerbepath. Gewerbehyg. 9:477–486, 1939.
34. Baader, E. W. Erkrankungen durch Mangan. Osaka City Med. J. 5:121–128, 1959.
35. Baader, E. W. Manganpneumonie. Aerstl. Sachverst.-Zeitung 43:75–81, 1937.
36. Bach, S. J., and D. B. Whitehouse. Purification and properties of arginase. (Proceedings of the Biochemical Society–Abstract.) Biochem. J. 57:xxxi, 1954.
37. Baker, D. E., B. R. Bradford, and W. I. Thomas. Leaf analysis of corn–tool for predicting soil fertility needs. Better Crops Plant Food 50:36–40, 1966.
38. Balani, S. G., G. M. Umarji, R. A. Bellare, and H. C. Merchant. Chronic manganese poisoning. A case report. J. Postgrad. Med. 13:116–122, 1967.
39. Barbeau, A. The pathogenesis of Parkinson's disease: A new hypothesis. Can. Med. Assoc. J. 87:802–807, 1962.
40. Barbeau, A., G. F. Murphy, and T. S. Sourkes. Excretion of dopamine in diseases of the basal ganglia. Science 133:1706–1707, 1961.
41. Barnard, P. G., A. G. Starliper, W. M. Dressel, and M. M. Fine. Recycling of Steelmaking Dusts. U.S. Bureau of Mines Solid Waste Program Technical Progress Report 52. Pittsburgh: U.S. Department of the Interior, Bureau of Mines, 1972. 10 pp.
42. Barnard, P. G., D. F. Walsh, and J. A. Rowland. Problems in Substituting Titanium for Manganese in Steel. U.S. Bureau of Mines Report of Investigations 6030. Washington, D.C.: U.S. Department of the Interior, 1962. 27 pp.
43. Barnes, L. L., G. Sperling, and L. A. Maynard. Bone development in the albino rat on a low manganese diet. Proc. Soc. Exp. Biol. Med. 46:562–565, 1941.
44. Bartels, T. T., and C. E. Wilson. Determination of methyl cyclopentadienyl manganese tricarbonyl in JP-4 fuel by atomic absorption spectrophotometry. Atomic Absorption Newslett. 8:3–5, 1969.

45. Bear, F. E. Chemistry of the Soil. (2nd ed.) (American Chemical Society Monograph Series No. 160) New York: Reinhold Publishing Corp., 1964. 515 pp.
46. Bear, F. E., and A. Wallace. Alfalfa. Its Mineral Requirements and Chemical Composition. New Jersey Agricultural Experiment Station Bulletin 748. New Brunswick: Rutgers, The State University, 1950. 32 pp.
47. Bellare, R. A. Studies in manganese poisoning. Ph.D. thesis, University of Bombay (India), 1967. 84 pp. (typescript)
48. Bentley, O. G., and P. H. Phillips. The effect of low manganese rations upon dairy cattle. J. Dairy Sci. 34:396–403, 1951.
49. Berg, W. A., and W. G. Vogel. Manganese Toxicity of Legumes Seeded in Kentucky Strip-Mine Spoils. U.S. Forest Service Research Paper NE-119. Upper Darby, Pa.: U.S. Department of Agriculture Northeastern Forest Experimental Station, 1968. 12 pp.
50. Berger, K. C., and G. C. Gerloff. Manganese toxicity of potatoes in relation to strong soil acidity. Soil Sci. Soc. Amer. Proc. 12:310–314, 1947.
51. Berger, K. C., and P. F. Pratt. Advances in secondary and micronutrient fertilization, pp. 287–340. In M. H. McVickar, G. L. Bridger, and L. B. Nelson, Eds. Fertilizer Technology and Usage. Proceedings of a Short Course sponsored by the Soil Science Society of America and held at Purdue University, Lafayette, Indiana, February 12–13, 1962. Madison, Wisc.: Soil Science Society of America, 1963.
52. Bernard, R., and J. M. Demers. Le manganèse dans la nutrition du caneton (Pékin blanc). Rev. Can. Biol. 11:147–157, 1953.
53. Bernheimer, H., W. Birkmayer, and O. Hornykiewicz. Zur biochemie des Parkinson-Syndroms des Menschen. Einfluss der Monoaminoxydase-Hemmer-Therapie auf Konzentration des Dopamins Noradrenalins und 5-Hydroxytryptamins im Gehirn. Klin. Wochenschr. 41:465–469, 1963.
54. Bernheimer, H., and O. Hornykiewicz. Das Verhalten des Dopamin-Metaboliten Homovanillinsäure im Gehirn von normalen und Parkinson-Kranken Menschen. Naunyn Schmiedebergs Arch. Pharmakol. 247:305–306, 1964.
55. Bernheimer, H., and O. Hornykiewicz. Herabgesetzte Konzentration der Homovanillinsäure im Gehirn von Parkinson-Kranken Menschen als Ausdruck der Störung des zentralen Dopaminstoffwechsels. Klin. Wochenschr. 43:711–715, 1965.
56. Bertinchamps, A. J., and G. C. Cotzias. Practical ultrafiltration of transition metals in plasma. Fed. Proc. 18:469, 1959. (abstract)
57. Bertinchamps, A. J., S. T. Miller, and G. C. Cotzias. Interdependence of routes excreting manganese. Amer. J. Physiol. 211:217–224, 1966.
58. Bethlehem Steel Corporation. Laboratory Procedures. Manganese Biological Sampling and Analysis. Environmental Health Chemistry Section, Environmental Quality Control Div., Industrial Relations Dept. (personal communication, 1972)
59. Bhagat, S. K., W. H. Funk, H. F. Royston, and K. R. Shah. Trace element analysis of environmental samples by neutron activation method. J. Water Pollut. Control Fed. 43:2414–2423, 1971.
60. Bishop, C. A., A. H. Brisse, R. G. Thompson, C. R. Liebel, and W. A. Swaney. Cleaning ferromanganese blast furnace gas. Iron Steel Eng. 28:134–136, 1951.

61. Bishop, C. A., W. W. Campbell, D. L. Hunter, and M. W. Lightner. Successful cleaning of open-hearth exhaust gas with a high-energy Venturi scrubber. J. Air Pollut. Control Assoc. 11:83–87, 1961.
62. Böckmann, P. W. K. pp. 2548–2550. In Discussion of a paper by D. Elstad, Factory smoke containing manganese as contributing cause in pneumonia epidemics in an industrial district. Nord. Med. 3:2527–2533, 1939. (in Norwegian)
63. Bollard, E. G. Manganese deficiency of apricots. New Zealand J. Sci. Tech. 34-A:471–472, 1953.
64. Borg, D. C., and G. C. Cotzias. Interaction of trace metals with phenothiazine drug derivatives. I. Structure–reactivity correlations. Proc. Natl. Acad. Sci. U.S.A. 48:617–623, 1962.
65. Borg, D. C., and G. C. Cotzias. Interaction of trace metals with phenothiazine drug derivatives. II. Formation of free radicals. Proc. Natl. Acad. Sci. U.S.A. 48:623–642, 1962.
66. Borg, D. C., and G. C. Cotzias. Manganese metabolism in man: Rapid exchange of Mn^{56} with tissue as demonstrated by blood clearance and liver uptake. J. Clin. Invest. 37:1269–1278, 1958.
67. Borsook, H., and J. W. Dubnoff. On the role of the oxidation in the methylation of guanidoacetic acid. J. Biol. Chem. 171:363–375, 1947.
68. Bortner, C. E. Toxicity of manganese to Turkish tobacco in acid Kentucky soils. Soil Sci. 39:15–33, 1935.
69. Bowen, H. J. M. Trace Elements in Biochemistry. New York: Academic Press, Inc., 1966. 241 pp.
70. Boyer, J., and J. Rodier. Aspects neurologiques de l'intoxication professionnelle par le manganese. Rev. Neurol. 90:13–27, 1954.
71. Boyer, P. D., J. H. Shaw, and P. H. Phillips. Studies on manganese deficiency in the rat. J. Biol. Chem. 143:417–425, 1942.
72. Braidech, M. M., and F. H. Emery. The spectrographic determination of minor chemical constituents in various water supplies in the United States. J. Amer. Water Works Assoc. 27:557–580, 1935.
73. Bremner, I., and A. H. Knight. The complexes of zinc, copper, and manganese present in ryegrass. Brit. J. Nutr. 24:279–290, 1970.
74. Brezina, E. Internationale Übersicht über Gewerbekrankheiten nach den Berichten der Gewerbeinspektionen der Kulturländer über das Jahr 1913. Vol. 8. Berlin: J. Springer, 1921. 143 pp.
75. Brief, R. S., J. W. Blanchard, R. A. Scala, and J. H. Blacker. Metal carbonyls in the petroleum industry. Arch. Environ. Health 23:373–384, 1971.
76. Brief, R. S., A. H. Rose, Jr., and D. G. Stephan. Properties and control of electric-arc steel furnace fumes. J. Air Pollut. Control Assoc. 6:220–224, 1957.
77. Britton, A. A., and G. C. Cotzias. Dependence of manganese turnover on intake. Amer. J. Physiol. 211:203–206, 1966.
78. Bromfield, S. M. The properties of a biologically formed manganese oxide, its availability to oats, and its solution by root workings. Plant Soil 9:325–337, 1958.
79. Brückmann, G., and S. G. Zondek. Iron, copper and manganese in human organs at various ages. Biochem. J. 33:1845–1857, 1939.

80. Bryant, D. R., and A. R. Hodges. Burner memory following atomic absorption spectrophotometric analysis of moderately volatile organometallic complexes. Anal. Chem. 44:405–407, 1972.
81. Bubarev, A. Gigiena Bezopnnost i Patologiia truda, 1931.
82. Burnett, W. T., Jr., R. R. Bigelow, A. W. Kimball, and C. W. Sheppard. Radiomanganese studies on the mouse, rat and pancreatic fistula dog. Amer. J. Physiol. 168:620–625, 1952.
83. Butt, E. M., R. E. Nusbaum, T. C. Gilmour, S. L. Didio, and Sister Mariano. Trace metal levels in human serum and blood. Arch. Environ. Health 8:52–57, 1964.
84. Büttner, H. E., and E. Lenz. Über die Möglichkeit von Manganschäden im Braunsteinbergwerk. Arch. Gewerbepath. Gewerbehyg. 7:672–684, 1937.
85. Cannon, H. L., and B. M. Anderson. The geochemist's involvement with the pollution problem, pp. 155–177. In H. L. Cannon and H. C. Hopps, Eds. Environmental Geochemistry in Health and Disease. American Association for Advancement of Science Symposium, Dallas, Texas, December 1968. Geological Society of America, Inc., Memoir 123. Boulder, Col.: The Geological Society of America, Inc., 1971.
86. Cantoni, G. L. Activation of methionine for transmethylation. J. Biol. Chem. 189:745–754, 1951.
87. Cantoni, G. L. Methylation of nicotinamide with a soluble enzyme system from rat liver. J. Biol. Chem. 189:203–216, 1951.
88. Carlberg, J. R., J. V. Crable, L. P. Limtiaca, H. B. Norris, J. L. Holtz, P. Mauer, and F. R. Wolowicz. Total dust, coal, free silica and trace metal concentrations in bituminous coal miners' lungs. Amer. Ind. Hyg. Assoc. J. 32:432–440, 1971.
89. Caro, J. H. Characterization of superphosphate, pp. 273–305. In U.S. Department of Agriculture and Tennessee Valley Authority. Superphosphate—Its History, Chemistry, and Manufacture. Washington, D.C.: U.S. Department of Agriculture, 1964.
90. Caskey, C. D., W. D. Gallup, and L. C. Norris. The need for manganese in the bone development of the chick. J. Nutr. 17:407–417, 1939.
91. Caskey, C. D., and L. C. Norris. Micromelia in adult fowl caused by manganese deficiency during embryonic development. Proc. Soc. Exp. Biol. Med. 44:332–335, 1940.
92. Cheng, B. T., and G. J. Ouellette. Effect of various anions on manganese toxicity in *Solanum tuberosum*. Can. J. Soil Sci. 48:109–115, 1968.
93. Cheng, B. T., and G. J. Ouellette. Effects of steam sterilization and organic amendments on the manganese status and associated characteristics of acid soils. Soil Sci. 110:383–388, 1970.
94. Cho, D. Y., and F. N. Ponnamperuma. Influence of soil temperature on the chemical kinetics of flooded soils and the growth of rice. Soil Sci. 112:184–194, 1971.
95. Cholak, J., and D. M. Hubbard. Determination of manganese in air and biological material. Amer. Ind. Hyg. Assoc. J. 21:356–360, 1960.
96. Clark, L. J., and W. L. Hill. Occurrences of manganese, copper, zinc, molybdenum, and cobalt in phosphate fertilizers and sewage sludge. J. Assoc. Offic. Agric. Chem. 41:631–637, 1958.
97. Clarke, F. W. The Data of Geochemistry, pp. 175–211. U.S. Geological Survey Bulletin 695. Washington, D.C.: U.S. Government Printing Office, 1920.

References

98. Clarke, F. W., and H. S. Washington. The composition of the earth's crust, pp. 12–44. In U.S. Geological Survey Professional Paper 127. Washington, D.C.: U.S. Government Printing Office, 1924.
99. Cohn, M. Magnetic resonance studies of metal activation of enzymic reactions of nucleotides and other phosphate substrates. Biochemistry 2:623–629, 1963.
100. Collander, R. Selective absorption of cations by higher plants. Plant Physiol. 16:691–720, 1941.
101. Collins, J. F., and S. W. Buol. Effects of fluctuation in the Eh-pH environment of iron and/or manganese equilibria. Soil Sci. 110:111–118, 1970.
102. Coorts, G. D. Excess Manganese Nutrition of Ornamental Plants. Missouri Agricultural Experiment Station Research Bulletin 669. Columbia: University of Missouri, 1958. 35 pp.
103. Coppenet, M., and J. Calvez. Observation d'un cas d'intoxication manganique de la pomme de terre sur sol très acide. C. R. Acad. Agric. France 46:728–734, 1960.
104. Cotzias, G. C. Importance of trace substances in environmental health as exemplified by manganese, pp. 5–19. In Proceedings, University of Missouri's 1st Annual Conference on Trace Substances in Environmental Health. July 10–11, 1967, Columbia, Missouri. Columbia: University of Missouri.
105. Cotzias, G. C. Manganese, pp. 403–442. In C. L. Comar and F. Bronner, Eds. Mineral Metabolism: An Advanced Treatise. Vol. II. The Elements. Part B. New York: Academic Press, 1962.
106. Cotzias, G. C. Manganese in health and disease. Physiol. Rev. 38:503–532, 1958.
107. Cotzias, G. C. Manganese, melanins and the extrapyramidal system. J. Neurosurg. 24 (Suppl.):170–180, 1966.
108. Cotzias, G. C. Metabolic modification of some neurologic disorders. J.A.M.A. 210:1255–1262, 1969.
109. Cotzias, G. C., and V. P. Dole. Metabolism of amines. II. Mitochondrial localization of monoamine oxidase. Proc. Soc. Exp. Biol. Med. 78:157–160, 1951.
110. Cotzias, G. C., and P. S. Papavasiliou. State of binding of natural manganese in human cerebrospinal fluid, blood and plasma. Nature 195:823–824, 1962.
111. Cotzias, G. C., P. S. Papavasiliou, and R. Gellene. Modification of Parkinsonism—chronic treatment with L-dopa. New Eng. J. Med. 280:337–345, 1969.
112. Cotzias, G. C., P. S. Papavasiliou, J. Ginos, A. Steck, and S. Düby. Metabolic modification of Parkinson's disease and of chronic manganese poisoning. Ann. Rev. Med. 22:305–326, 1971.
113. Cotzias, G. C., P. S. Papavasiliou, E. R. Hughes, L. Tang, and D. C. Borg. Slow turnover of manganese in active rheumatoid arthritis accelerated by prednisone. J. Clin. Invest. 47:992–1001, 1968.
114. Cotzias, G. C., P. S. Papavasiliou, and S. T. Miller. Manganese in melanin. Nature 201:1228–1229, 1964.
115. Cotzias, G. C., P. S. Papavasiliou, M. H. Van Woert, and A. Sakamato. Melanogenesis and extrapyramidal diseases. Fed. Proc. 23:713–718, 1964.
116. Coughanowr, D. R., and F. E. Krause. The reaction of SO_2 and O_2 in aqueous solutions of $MnSO_4$. Ind. Eng. Chem. Fundamentals 4:61–66, 1965.

117. Coulter, R. S. Smoke, dust, fumes—closely controlled in electric furnaces. Iron Age 173:107–110, Jan. 14, 1954.
118. Couper, J. On the effects of black oxide of manganese when inhaled into the lungs. Brit. Ann. Med. Pharm. Vit. Stat. Gen. Sci. 1:41–42, 1837.
119. Cox, F. R. Development of a yield response prediction and manganese soil test interpretation for soybeans. Agron. J. 60:521–524, 1968.
120. Cunningham, G. N., M. B. Wise, and E. R. Barrick. Effect of high dietary levels of manganese on the performance and blood constituents of calves. J. Anim. Sci. 25:532–538, 1966.
121. Curran, G. L. Effect of certain transition group elements on hepatic synthesis of cholesterol in the rat. J. Biol. Chem. 210:765–770, 1954.
122. Czechoslovak Committee of MAC. Manganese, pp. 106–107. In Documentation of MAC in Czechoslovakia. Prague: Czechoslovakia Ministry of Health, 1969.
123. Dählstrom, A., and K. Fuxe. Evidence for the existence of monoamine-containing neurons in the central nervous system. Acta Physiol. Scand. 62 (Suppl. 232):1–55, 1964.
124. Dams, R., J. A. Robbins, K. A. Rahn, and J. W. Winchester. Nondestructive neutron activation analysis of air pollution particulates. Anal. Chem. 42:861–867, 1970.
125. Dantin Gallego, J. Higiene y Patologia del Trabajo con Manganeso. Instituto Nacional de Prevision Publication 445. Madrid: Instituto Nacional de Prevision, 1935.
126. Dantin Gallego, J. Intoxicacion por el manganeso, pp. 459–473. In Curso de Higiene del Trabajo. Madrid: Publicaciones de la Jeffatura Provincial de Sanidad, 1944.
127. DeHuff, G. L. Manganese, pp. 493–509. In U.S. Bureau of Mines. Mineral Facts and Problems. (1956 ed.) U.S. Bureau of Mines Bulletin 556. Washington, D.C.: U.S. Government Printing Office, 1956.
128. DeHuff, G. L. Manganese, pp. 493–510. In U.S. Bureau of Mines. Mineral Facts and Problems. (1960 ed.) U.S. Bureau of Mines Bulletin 585. Washington, D.C.: U.S. Government Printing Office, 1960.
129. DeHuff, G. L. Manganese, pp. 553–572. In U.S. Bureau of Mines. Mineral Facts and Problems. (1965 ed.) U.S. Bureau of Mines Bulletin 630. Washington, D.C.: U.S. Government Printing Office, 1965.
130. DeHuff, G. L. Manganese, pp. 691–703. In 1970 Bureau of Mines Minerals Yearbook. Washington, D.C.: U.S. Government Printing Office, 1972.
131. Denton, M. D., and A. Ginsburg. Conformational changes in glutamine synthetase from Escherichia coli. I. The binding of Mn^{2+} in relation to some aspects of the enzyme structure and activity. Biochemistry 8:1714–1725, 1969.
132. Dervillée, P., G. Morichaud-Beuchant, M.-J. Charpentier, and E. Dervillée. A propos de la pneumonie manganique. Arch. Mal. Prof. 27:222–224, 1966.
133. Dessureaux, L. Heritability of tolerance to manganese toxicity in lucerne. Euphytica 8:260–265, 1959.
134. Dessureaux, L. The reaction of lucerne seedlings to high concentrations of manganese. Plant Soil 13:114–122, 1960.
135. Dessureaux, L., and G. J. Ouellette. Tolerance of alfalfa to manganese toxicity in sand culture. Can. J. Soil Sci. 38:8–13, 1958.
136. Dogan, S., and T. Beritic. Industrial and clinical aspects of occupational

poisoning with manganese. Arhiv za Higijena Rada (Zagreb) 4:139–212, 1953. (in Serbo-Croatian)
137. Doi, Y. Studies on the oxidizing power of roots of crop plants. I. The difference with species of crop plants and wild grasses. Proc. Crop Sci. Soc. Japan 21:12–13, 1952. (in Japanese; summary in English)
138. Doi, Y. Studies on the oxidizing power of roots of crop plants. II. The interrelation between paddy rice and soy-bean. Proc. Crop Sci. Soc. Japan 21:14–15, 1952. (in Japanese; summary in English)
139. Dokiya, Y., N. Owa, and S. Mitsui. Comparative physiological study of iron, manganese and copper absorption by plants. III. Interaction of Fe, Mn, and Cu on the absorption of the elements by rice and barley seedlings. Soil Sci. Plant Nutr. 14:169–174, 1968.
140. Dokuchaev, V. F., and N. N. Skvortsova. Atmospheric air pollution with manganese compounds and their effect on the organism, pp. 40–46. In U.S.S.R. Literature on Air Pollution and Related Occupational Diseases. Vol. 9. Part 1. Translated by B. S. Levine. Moscow: Meditsina Press, 1962. (Available from National Technical Information Service, Springfield, Va., as report TT 64-11574)
141. Dokuchaeva, V. F., and N. N. Skvortsova. Biological effect of experimental inhalation of low manganese concentrations. U.S.S.R. Literature on Air Pollution and Related Occupational Diseases 16:98–104, 1966.
142. Drosdoff, M. Leaf composition in relation to the mineral nutrition of tung trees. Soil Sci. 57:281–291, 1944.
143. Duffy, P. E., and V. M. Tennyson. Phase and electron microscopic observations of Lewy bodies and melanin granules in the substantia nigra and locus caeruleus in Parkinson's disease. J. Neuropath. Exp. Neurol. 24:398–414, 1965.
144. Durfor, C. N., and E. Becker. Public Water Supplies of the 100 Largest Cities in the United States, 1962. U.S. Geological Survey, Water-Supply Paper 1812. Washington, D.C.: U.S. Government Printing Office, 1964. 364 pp.
145. Durum, W. H., and J. Haffty. Implications of the minor element content of some major streams of the world. Geochim. Cosmochim. Acta 27:1–11, 1963.
146. Durum, W. H., and J. Haffty. Occurrence of Minor Elements in Water. U.S. Geological Survey Circular 445. Washington, D.C.: U.S. Geological Survey, 1961. 11 pp.
147. Dykstra, F. R. Manganese chemicals and their uses. Industrial Minerals (London) No. 30, pp. 15–17, March 1970.
148. Ehringer, H., and O. Hornykiewicz. Verteilung von Noradrenalin und Dopamin (3-Hydroxytyramin) im Gehirn des Menschen und ihr Verhalten bei Erkrankungen des extrapyramidalen Systems. Klin. Wochenschr. 38:1236–1239, 1960.
149. Ellis, G. H., S. E. Smith, E. M. Gates, D. Lobb, and E. J. Larson. Further studies of manganese deficiency in the rabbit. J. Nutr. 34:21–31, 1947.
150. Elstad, D. Beobachtungen über Manganpneumonien, Vol. 2, pp. 1014–1022. In Bericht über den VIII. Internationalen Kongress für Unfallmedizin und Berufskrankheiten. Frankfurt A.M., 26–28 September 1938. (2 vols.) Leipzig: Thieme, 1939.
151. Elstad, D. Factory smoke containing manganese as contributing cause in

pneumonia epidemics in an industrial district. Nord. Med. 3:2527-2533, 1939. (in Norwegian)
152. Emden, H. Ueber eine Nervenkrankheit nach Manganvergiftung. Munch. Med. Wochenschr. 48:1852-1853, 1901.
153. Emmel, M. W. Perosis in swans and chickens fed manganese-fortified mashes. J. Amer. Vet. Med. Assoc. 104:30, 32, 1944.
154. Epstein, E. Effect of soil temperature on mineral element composition and morphology of the potato plant. Agron. J. 63:664-666, 1971.
155. Epstein, E., and O. Lilleland. A preliminary study of the manganese content of the leaves of some deciduous fruit trees. Proc. Amer. Soc. Hort. Sci. 41: 11-18, 1942.
156. Erickson, O. E. Dust control of electric steel foundries in Los Angeles area, pp. 157-160 (discussion 160-163). In Electric Furnace Steel Proceedings, 1953. Vol. 11. New York: American Institute of Mining and Metallurgical Engineers, 1954.
157. Erway, L., L. S. Hurley, and A. Fraser. Neurological defect: Manganese in phenocopy and prevention of a genetic abnormality of inner ear. Science 152:1766-1768, 1966.
158. Erway, L., L. S. Hurley, and A. S. Fraser. Congenital ataxia and otolith defects due to manganese deficiency in mice. J. Nutr. 100:643-654, 1970.
159. Ethyl Corporation. The Effect of Manganese on the Oxidation of SO_2. Detroit, Mich.: Ethyl Corporation, 1971. 5 pp.
160. Evans, R. J., E. I. Robertson, M. Rhian, and L. A. Wilhelm. The development of perosis in turkey poults and its prevention. Poult. Sci. 21:422-429, 1942.
161. Everson, G. J., L. S. Hurley, and J. F. Geiger. Manganese deficiency in the guinea pig. J. Nutr. 68:49-56, 1959.
162. Everson, G. J., and R. E. Shrader. Abnormal glucose tolerance in manganese-deficient guinea pigs. J. Nutr. 94:89-94, 1968.
163. Fairhall, L. T., and P. A. Neal. Industrial Manganese Poisoning. National Institute of Health Bulletin 182. Washington, D.C.: U.S. Government Printing Office, 1943. 24 pp.
164. Fergus, I. F. Manganese toxicity in an acid soil. Queensland J. Agric. Anim. Sci. 11:15-27, 1954.
165. Ferrari, R. Experiences in developing an effective pollution control system for a submerged arc ferroalloy furnace operation. (Presented at the 25th Electric Furnace Conference, 1967) J. Metals 20:95-104, 1968.
166. Ferree, D. C., and A. H. Thompson. Internal Bark Necrosis of the Apple as Influenced by Calcium Placement and Soil Manganese. Maryland Agricultural Experiment Station Bulletin A-166. College Park: University of Maryland, 1970. 51 pp.
167. Finn, R. J., S. J. Bourget, K. F. Nielsen, and B. K. Dow. Effects of different soil moisture tensions on grass and legume species. Can. J. Soil Sci. 41:16-23, 1961.
168. Fishman, M. J., and D. E. Erdmann. Water analysis. (Analytical Review, 1971 Applications) Anal. Chem. 43(5):365R-388R, 1971.
169. Flawn, P. T. Environmental Geology: Conservation, Land-Use Planning, and Resource Management, p. 131. New York: Harper and Row, 1970.
170. Flinn, R. H., P. A. Neal, and W. B. Fulton. Industrial manganese poisoning. J. Ind. Hyg. Toxicol. 23:374-387, 1941.
171. Flinn, R. H., P. A. Neal, W. H. Reinhart, J. M. Dallavalle, W. B. Fulton, and

References

A. D. Dooley. Chronic Manganese Poisoning in an Ore-Crushing Mill. (Public Health Bulletin 247, Federal Security Agency) Washington, D.C.: U.S. Government Printing Office, 1940. 77 pp.

172. Follett, R. H., and W. L. Lindsay. Changes in DTPA-extractable zinc, iron, manganese, and copper in soils following fertilization. Soil Sci. Soc. Amer. Proc. 35:600–602, 1971.

173. Foltz, A. K., J. A. Yeransian, and K. G. Sloman. Food. (Analytical Reviews, 1971 Applications) Anal. Chem. 43(5):70R–100R, 1971.

174. Foradori, A. C., A. Bertinchamps, J. M. Gulibon, and G. C. Cotzias. The discrimination between magnesium and manganese by serum proteins. J. Gen. Physiol. 50:2255–2266, 1967.

175. Fore, H., and R. A. Morton. Microdetermination of manganese in biological material by a modified catalytic method. Biochem. J. 51:594–598, 1952.

176. Foy, C. D. Characterization of Manganese, Aluminum, and Hydrogen Ion Toxicities in Alfalfa. Research Report 365, USDA, ARS, SWC, U.S. Soils Laboratory, Beltsville, Maryland. (in-house publication available on request; unpublished research report)

177. Foy, C. D. Toxic Factors in Acid Soils of the Southwestern United States as Related to the Response of Alfalfa to Lime. U.S. Department of Agriculture Production Research Report 80. Washington, D.C.: U.S. Government Printing Office, 1964. 26 pp.

178. Foy, C. D., A. L. Fleming, and W. H. Armiger. Differential tolerance of cotton varieties to excess manganese. Agron. J. 61:690–694, 1969.

179. Foy, C. D., A. L. Fleming, and J. W. Schwartz. Opposite aluminum and manganese tolerances of two wheat varieties. Agron. J. 65:123–126, 1973.

180. Freise, F. W. Gesundheitsschädigungen durch Verarbeitung pflanzlicher Drogen. Beobachtungen aus brasilianischen Gewerbebetrieben. Arch. Gewerbepath. Gewerbehyg. 4:381–399, 1933.

181. Fried, J. F., A. Lindenbaum, and J. Schubert. Comparison of 3 chelating agents in treatment of experimental manganese poisoning. Proc. Soc. Exp. Biol. Med. 100:570–573, 1959.

182. Friedmann, N., and H. Rasmussen. Calcium, manganese and hepatic gluconeogenesis. Biochim. Biophys. Acta 222:41–52, 1970.

183. Fuenzalida, P., and I. Mena. Intoxicacion cronica por manganeso y sus relaciones con las enfermedades del sistema extrapiramidal. Rev. Med. Chile 95:667–674, 1965.

184. Fujimoto, C. K., and G. D. Sherman. The effect of drying, heating and wetting on the level of exchangeable manganese in Hawaiian soils. Soil Sci. Soc. Amer. Proc. 10:107–112, 1945.

185. Fujiwara, A., and H. Ishida. Acceleration of manganese uptake by rice plant grown under unfavorable temperature or light condition. Tohoku J. Agric. Res. 14:209–215, 1964.

186. Gallagher, P. H., and T. Walsh. The susceptibility of cereal varieties to manganese deficiency. J. Agric. Sci. 33:197–203, 1943.

187. Gallup, W. D., and L. C. Norris. The amount of manganese required to prevent perosis in the chick. Poult. Sci. 18:76–82, 1939.

188. Garcia Avila, M., and R. Penalver Ballina. Manganese poisoning in the mines of Cuba. A preliminary report. Ind. Med. Surg. 22:220–221, 1953.

189. Gaw, R. G. Gas cleaning. Iron Steel Eng. 37:81–85, 1960.

190. Geering, H. R., J. F. Hodgson, and C. Sdano. Micronutrient cation com-

plexes in soil solution. IV. The chemical state of manganese in soil solution. Soil Sci. Soc. Amer. Proc. 33:81-85, 1969.
191. Gerhard, E. R., and H. F. Johnstone. Photochemical oxidation of sulfur dioxide in air. Indust. Eng. Chem. 47:972-976, 1955.
192. Gerloff, G. C., D. G. Moore, and J. T. Curtis. Selective absorption of mineral elements by native plants of Wisconsin. Plant Soil 25:393-405, 1966.
193. Gerloff, G. C., P. R. Stout, and L. H. P. Jones. Molybdenum-manganese-iron antagonisms in the nutrition of tomato plants. Plant Physiol. 34:608-613, 1959.
194. Gerstle, R. W., S. T. Cuffe, A. A. Orning, and C. H. Schwartz. Air pollutant emissions from coal-fired power plants. Report No. 2. J. Air Pollut. Control Assoc. 15:59-64, 1965.
195. Gheesling, R. H., and H. F. Perkins. Critical levels of manganese and magnesium in cotton at different stages of growth. Agron. J. 62:29-32, 1970.
196. Gilbert, B. E., F. T. McLean, and L. J. Hardin. The relation of manganese and iron to a lime-induced chlorosis. Soil Sci. 22:437-446, 1926.
197. Gillert, O. Galvanischer Strom—Faradischer Strom—Exponentialstrom in der therapeutischen Praxis. Munich: Pflaum, 1953-1954. 80 pp.
198. Gladstones, J. S., and D. P. Drover. The mineral composition of lupins. 1. A survey of the copper, molybdenum, and manganese contents of lupins in the south west of Western Australia. Aust. J. Exper. Agric. Anim. Husb. 2:46-53, 1962.
199. Glassman, E. The biochemistry of learning: an evaluation of the role of RNA and protein. Ann. Rev. Biochem. 38:605-646, 1969.
200. Goldberg, E. D. Minor elements in sea water, pp. 163-196. In J. P. Riley and G. Skirrow, Eds. Chemical Oceanography. Vol. 1. London and New York: Academic Press, Inc., 1965.
201. Goldberg, E. D., and G. Arrhenius. Chemistry of Pacific pelagic sediments. Geochim. Cosmochim. Acta 13:153-212, 1958.
202. Goldschmidt, V. M. The principles of distribution of chemical elements in minerals and rocks, pp. 655-673. In Journal of the Chemical Society, 1937 Proceedings, Part 1. London, 1937. (The Seventh Hugo Müller Lecture, Delivered before the Chemical Society on March 17, 1937)
203. Good, C. H., Jr. Ferromanganese furnace fumes cleaned successfully. Iron Age 174:95-97, July 8, 1954.
204. Gorsline, G. W., D. E. Baker, and W. I. Thomas. Accumulation of eleven elements by field corn (*Zea mays* L.), pp. 25-33. In Pennsylvania State University Bulletin 725. University Park: The Pennsylvania State University, 1965.
205. Grashuis, J., J. J. Lehr, L. O. E. Beuvery, and A. Beuvery-Asman. Manganese deficiency in cattle. Hoogland. Instituut voor Moderne Veevoeding "de Schothorst." Mededeling, No. 40:1-16, 1953. (in Dutch)
206. Graven, E. H., O. J. Attoe, and D. Smith. Effect of liming and flooding on manganese toxicity in alfalfa. Soil Sci. Soc. Amer. Proc. 29:702-706, 1965.
207. Greenberg, D. M., D. H. Copp, and E. M. Cuthbertson. Studies in mineral metabolism with the aid of artificial radioactive isotopes. VII. The distribution and excretion, particularly by way of the bile, of iron, cobalt, and manganese. J. Biol. Chem. 147:749-756, 1943.
208. Grummer, R. H., O. G. Bentley, P. H. Phillips, and G. Bohstedt. The role of manganese in growth, reproduction, and lactation of swine. J. Anim. Sci. 9:170-175, 1950.

References 161

209. Gupta, U. C., E. W. Chipman, and D. C. MacKay. Influence of manganese and pH on chemical composition, bronzing of leaves, and yield of carrots grown on acid sphagnum peat soil. Soil Sci. Soc. Amer. Proc. 34:762-764, 1970.
210. Habermann, H. M. Light-dependent oxygen metabolism of chloroplast preparations. Plant Physiol. 35:307-312, 1960.
211. Hale, J. B. Deficiency diseases of the sugar beet. Unpublished report, duplicated for private circulation. Agr. Res. Coun. 7828:8, 1945.
212. Hammes, J. K., and K. C. Berger. Chemical extraction and crop removal of manganese from air-dried and moist soils. Soil Sci. Soc. Amer. Proc. 24:361-364, 1960.
213. Hanna, W. J., and T. B. Hutcheson, Jr. Soil-plant relationships, pp. 141-162. In L. B. Nelson, Ed. Changing Patterns in Fertilizer Use. Proceedings of a symposium sponsored by the Soil Science Society of America, February 14-15, 1968 in Chicago. Madison, Wisc.: Soil Science Society of America, Inc., 1968.
214. Hartman, R. H., G. Matrone, and G. H. Wise. Effect of high dietary manganese on hemoglobin formation. J. Nutr. 57:429-439, 1955.
215. Hasler, A. Über die Manganbedürftigkeit einiger Gräserarten. Schweiz. Landw. Monatsh. 29:300-305, 1951.
216. Hatch, T. F. The role of permissible limits for hazardous airborne substances in the working environment in the prevention of occupational diseases. Bull. WHO 47:151-159, 1972.
217. Hatch, T. F., and P. Gross. Pulmonary Deposition and Retention of Inhaled Aerosols. New York: Academic Press, 1964. 192 pp.
218. Hawkins, G. E., Jr., G. H. Wise, G. Matrone, R. K. Waugh, and W. L. Lott. Manganese in the nutrition of young dairy cattle fed different levels of calcium and phosphorus. J. Dairy Sci. 38:536-547, 1955.
219. Hay, J. O. Manganese compounds, pp. 1-55. In R. E. Kirk and D. F. Othmer, Eds. Kirk-Othmer Encyclopedia of Chemical Technology. Vol. 13. (2nd ed.) New York: Interscience Publishers, 1967.
220. Healy, W. B. Treatment of a lime-induced manganese deficiency in peach trees. New Zealand J. Sci. Tech. 34-A:386-396, 1953.
221. Heine, W. Beobachtungen und experimentelle Untersuchungen über Manganvergiftungen und "Manganpneumonien." Z. Hyg. Infektionskrank. 125:3-76, 1943.
222. Heintz, J. G. Manganese-phosphate reactions in aqueous systems and the effects of application of monocalcium phosphate on the availability of manganese to oats in an alkaline fen soil. Plant Soil 29:407-423, 1968.
223. Heintze, S. G. Manganese deficiency in peas and other crops in relation to the availability of soil manganese. J. Agric. Sci. 36:227-238, 1946.
224. Heller, V. G., and R. Penquite. Factors producing and preventing perosis in chickens. Poult. Sci. 16:243-246, 1937.
225. Hemeon, W. C. L., Ed. Air pollution problems of the steel industry. Technical Coordinating Committee T-6 Steel Report. J. Air Pollut. Control Assoc. 7:62-67, 1957.
226. Hewett, D. F. Manganese in sediments, pp. 562-581. In W. H. Twenhofel, Ed. Treatise on Sedimentation. Baltimore, Md.: Williams and Wilkins Co., 1932.
227. Hewett, D. F., M. D. Crittenden, L. Pavlides, and G. L. DeHuff. Manganese deposits of the United States, pp. 169-230. In Tomo III, America del Norte.

Symposium Sobre Yacimientos de Manganeso, XX Congreso Geologico Internacional, Mexico City, 1956.
228. Hewitt, E. J. Relation of manganese and some other metals to the iron status of plants. Nature 161:489–490, 1948.
229. Hewitt, E. J. The resolution of the factors in soil acidity. IV. The relative effects of aluminum and manganese toxicities on some farm and market garden crops. Ann. Report Agric. Hort. Res. Sta. 1948:58–65, 1948.
230. Hiatt, A. J., and J. L. Ragland. Manganese toxicity of burley tobacco. Agron. J. 55:47–49, 1963.
231. Hignett, S. L. The influence of calcium, phosphorus, manganese and Vitamin D on heifer fertility, pp. 116–123. In III International Congress on Animal Reproduction, 25–30 June 1956, Cambridge. London and Tonbridge: Brown Knight and Truscott, 1956.
232. Hill, R. M., D. E. Holtkamp, A. R. Buchanan, and E. K. Rutledge. Manganese deficiency in rats with relation to ataxia and loss of equilibrium. J. Nutr. 41:359–371, 1950.
233. Hines, C. R., and T. R. Dulski. Ferrous metallurgy. (Analytical Reviews, 1971 Applications) Anal. Chem. 43(5):100R–108R, 1971.
234. Hirsch-Kolb, H., and D. M. Greenberg. Molecular characteristics of rat liver arginase. J. Biol. Chem. 243:6123–6129, 1968.
235. Hoff, D. J., and H. J. Mederski. The chemical estimation of plant available soil manganese. Soil Sci. Soc. Amer. Proc. 22:129–137, 1958.
236. Höfner, W. Eisen- und manganhaltige Verbindungen im Blutungssaft von *Helianthus annuus*. Physiol. Plant. 23:673–677, 1970.
237. Holtkamp, D. E., and R. M. Hill. Effect on growth of the level of manganese in the diet of rats, with some observations on the manganese–thiamine relationship. J. Nutr. 41:307–316, 1950.
238. Hopkins, E. F., V. Pagán, and F. J. Ramirez Silva. Iron and manganese in relation to plant growth and its importance in Puerto Rico. J. Agric. Univ. Puerto Rico 28:43–101, 1944.
239. Hornykiewicz, O. Die topische Lokalisation und das Verhalten von Noradrenalin und Dopamin (3-Hydroxytyramin) in *der substantia nigra* des normalen und Parkinson-kranken Menschen. Wien. Klin. Wochenschr. 75:309–312, 1963.
240. Hornykiewicz, O. Dopamine (3-hydroxytyramine) and brain function. Pharmacol. Rev. 18:925–964, 1966.
241. Hornykiewicz, O. The role of brain dopamine (3-hydroxytyramine) in Parkinsonism, pp. 57–68. In E. Trabucchi, R. Paoletti, N. Canal, Eds., and L. Volicer, Ass't Ed. Biochemical and Neurophysiological Correlation of Centrally Acting Drugs. Proceedings of the Second International Pharmacological Meeting, August 20–23, 1963, Prague. Vol. 2. Praha: Czechoslovak Medical Press, 1964.
242. Howell, G. A. Air pollution control in the steel industry. Air Repair 3:163–166, 1954.
243. Hoyt, P. B., and M. Nyborg. Toxic metals in acid soil. II. Estimation of plant-available manganese. Soil Sci. Soc. Amer. Proc. 35:241–244, 1971.
244. Hurley, L. S. Genetic-nutritional interactions concerning manganese, pp. 41–51. In D. D. Hemphill, Ed. Trace Substances in Environmental Health. II. Proceedings of 2nd Annual Conference on Trace Substances in Environmental Health, July 16–18, 1968. Columbia: University of Missouri, 1968.

References

245. Hurley, L. S., and C. W. Asling. Localized epiphyseal dysplasia in offspring of manganese-deficient rats. Anat. Rec. 145:25–37, 1963.
246. Hurley, L. S., and G. J. Everson. Delayed development of righting reflexes in offspring of manganese-deficient rats. Proc. Soc. Exp. Biol. Med. 102:360–362, 1959.
247. Hurley, L. S., G. J. Everson, and J. F. Geiger. Manganese deficiency in rats: Congenital nature of ataxia. J. Nutr. 66: 309–319, 1958.
248. Hurley, L. S., G. J. Everson, and J. F. Geiger. Serum alkaline phosphatase activity in normal and manganese-deficient developing rats. J. Nutr. 67:445–450, 1959.
249. Hurley, L. S., D. E. Woolley, F. Rosenthal, and P. S. Timiras. Influence of manganese on susceptibility of rats to convulsions. Amer. J. Physiol. 204: 493–496, 1963.
250. Hurley, L. S., E. Wootten, G. J. Everson, and C. W. Asling. Anomalous development of ossification in the inner ear of offspring of manganese-deficient rats. J. Nutr. 71:15–19, 1960.
251. Hutchinson, G. E. A Treatise on Limnology. Vol. I. Geography, Physics, and Chemistry. New York: John Wiley and Sons, Inc., 1957. 1015 pp.
252. Hwang, J. Y., and F. J. Feldman. Determination of atmospheric trace elements by atomic absorption spectroscopy. Appl. Spectrosc. 24:371–374, 1970.
253. Ishizuka, Y., and T. Ondo. Interaction between manganese and zinc in growth of rice plants. Soil Sci. Plant Nutr. 14:201–206, 1968.
254. Iwakiri, B., and L. E. Scott. Mineral deficiency symptoms of the Temple strawberry grown in sand culture. Proc. Amer. Soc. Hort. Sci. 57:45–52, 1951.
255. Jackson, T. L., D. T. Westermann, and D. P. Moore. The effect of chloride and lime on the manganese uptake by bush beans and sweet corn. Soil Sci. Soc. Amer. Proc. 30:70–73, 1966.
256. Jackson, W. A. Physiological effects of soil acidity, pp. 43–124. In R. W. Pearson and F. Adams, Eds., R. C. Dinauer, Managing Ed. Soil Acidity and Liming. Number 12 in the series Agronomy. Madison, Wisc.: American Society of Agronomy, Inc., 1967.
257. Jacobson, H. G. M., and T. R. Swanback. Manganese content of certain Connecticut soils and its relation to the growth of tobacco. J. Amer. Soc. Agron. 24:237–245, 1932.
258. Jaksch, R. V. Cited in Prager Med. Wochenschr. 18:213–214, 1902.
259. Jenne, E. A. Controls on Mn, Fe, Co, Ni, Cu and Zn concentrations in soils and water: The significant role of hydrous Mn and Fe oxides. Adv. Chem. 73:337–387, 1968.
260. Joham, H. E., and J. V. Amin. The influence of foliar and substrate application of manganese on cotton. Plant Soil 26:369–379, 1967.
261. Johnson, S. R. Studies with swine on rations extremely low in manganese. J. Anim. Sci. 2:14–22, 1943.
262. Jones, H. H. Electrostatic precipitators, pp. B-4-1–B-4-3. In Air Sampling Instruments for Evaluation of Atmospheric Contaminants. (2nd ed.) Cincinnati: American Conference of Governmental Industrial Hygienists, 1962.
263. Jones, J. B., Jr. Distribution of fifteen elements in corn leaves. Soil Sci. Plant Anal. 1:27–33, 1970.

264. Jones, J. B., Jr. Soil and plant analysis as methods for diagnosing micronutrient deficiencies. Soil Sci. Plant Anal. 1:263-272, 1970.
265. Jones, L. H. P. The relative content of manganese in plants. Plant Soil 8:328-336, 1957.
266. Jones, W. W., and P. F. Smith. Nutrient deficiencies in citrus, pp. 359-414. In H. B. Sprague, Ed. Hunger Signs in Crops: A Symposium. (3rd ed.) New York: David McKay Co., 1964.
267. Jordine, C. G. Metal deficiencies in bananas. Nature 194:1160-1163, 1962.
268. Jötten, K. W., and H. Poppinga. Hygienische und experimentelle Studien über den Einfluss der Thomasschlackenstaubinhalation auf das Zustandekommen der Lungenentzündung. Arch. Hyg. Bakt. 115:61-74, 1935.
269. Jötten, K. W., and H. Reploh. Experimentelles zur Thomasschlackenstaubund Manganpneumonie, pp. 1028-1038. In Bericht über den VIII. Internationalen Kongress fur Unfallmedizin und Berufskrankheiten. Frankfurt A.M., 26-28 September 1938. (2 vols.) Leipzig: Thieme, 1939.
270. Jötten, K. W., H. Reploh, and G. Hegemann. Experimentelle Untersuchungen über die Manganpneumonie und ihre Beziehungen zur Thomasschlackenpneumonie. Arch. Gewerbepath. Gewerbehyg. 9:314-336, 1939.
271. Juday, C., E. A. Birge, and V. W. Meloche. Mineral content of the lake waters of northeastern Wisconsin. Wisconsin Acad. Sci. Arts Lett. Trans. 31:223-276, 1938.
272. Kamprath, E. J., and C. D. Foy. Lime-fertilizer-plant interactions in acid soils, pp. 105-151. In R. A. Olson, Ed. Fertilizer Technology and Use. (2nd ed.) Madison, Wisc.: Soil Science Society of America, Inc., 1971.
273. Katell, S., and K. D. Plants. Here's what SO_2 removal costs. Hydrocarbon Process. 46:161-164, 1967.
274. Kato, M. Distribution and excretion of radiomanganese administered to the mouse. Q. J. Exp. Physiol. 48:355-369, 1963.
275. Kawamura, R., H. Ikuta, S. Fukuzumi, R. Yamada, S. Tsubaki, T. Kodama, and S. Kurata. Intoxication by manganese in well water. Kitasato Arch. Exp. Med. 18:145-169, 1941.
276. Kemmerer, A. R., C. A. Elvehjem, and E. B. Hart. Studies on the relation of manganese to the nutrition of the mouse. J. Biol. Chem. 92:623-630, 1931.
277. Kent, N. L., and R. A. McCance. The absorption and excretion of "minor" elements by man; cobalt, nickel, tin and manganese. Biochem. J. 35:877-883, 1941.
278. Kesíc, B., and V. Häusler. Hematological investigation on workers exposed to manganese dust. A.M.A. Arch. Ind. Health Occup. Med. 10:336-343, 1954.
279. Khalatcheva, L., and V. L. Boyadjiev. Activité de la cystathionase chez des rats au cours de l'intoxication expérimentale par le manganese. J. Eur. Toxicol. 2:85-90, 1969.
280. Khavtasi, A. A. The possibility of chronic manganese poisoning among workers in manganese mines. Gig. Tr. Prof. Zabol. 2:36-39, 1958. (in Russian)
281. Khazan, G. L., Ya. M. Stanislavskiy, Yu. V. Vasilenko, L. V. Khizhnyakova, A. A. Baranenko, V. N. Kutepov, Z. F. Nestrugina, A. B. Nerubenko, and V. P. Protopopova. Working conditions and health status of workers in the Nikopol' manganese mines. Vrach. Delo Nos. 1-5:277-291, 1956. (in Russian)
282. Kinney, S. P., P. H. Royster, and T. L. Joseph. Iron Blast-Furnace Reactions.

References

U.S. Bureau of Mines Technical Paper 391. Washington, D.C.: U.S. Government Printing Office, 1927. 65 pp.

283. Kirsch, R. K., M. E. Harward, and R. G. Petersen. Interrelationships among iron, manganese, and molybdenum in the growth and nutrition of tomatoes grown in culture solution. Plant Soil 12:259–275, 1960.

284. Kleese, R. A., and L. J. Smith. Scion control of genotypic differences in mineral salts accumulated in soyabean (*Glycine max*. L. Merr.) seeds. Ann. Bot. N.S. 34:183–188, 1970.

285. Kleinkopf, M. D. Spectrographic determination of trace elements in lake waters of northern Maine. Geol. Soc. Amer. Bull. 71:1231–1242, 1960.

286. Kleinkopf, M. D. Trace element exploration of Maine lake water. Ph.D. dissertation, Columbia University, 1955. (University Microfilms Publication 12,447.) 157 pp.

287. Knezek, B. D., and J. F. Davis. Relative effectiveness of manganese sulfate and manganous oxide applied on organic soil. Soil Sci. Plant Anal. 2:17–21, 1971.

288. Konovalov, G. S. Removal of microelements by the principal rivers of the U.S.S.R. (Translated into English). Akad. Nauk SSSR Doklady 129:1034–1037, 1959. (Proceedings of Geological Society)

289. Konovalov, G. S., A. A. Ivanova, and T. K. Kolesnikova. Rare and dispersed elements (microelements) in the water and in the suspended substances in rivers of the European Territory of U.S.S.R. Gidrokhimicheskiye Materialy 42:94–111, 1966. (in Russian)

290. Kopp, J. F., and R. C. Kroner. Trace Metals in Waters of the United States. A Five-Year Summary of Trace Metals in Rivers and Lakes of the United States (Oct. 1, 1962–Sept. 30, 1967). Division of Pollution Surveillance, Federal Water Pollution Control Administration, U.S. Department of the Interior. Cincinnati: U.S. Department of the Interior, [1969]. 32 pp. and 16 appendixes.

291. Kosai, M. F., and A. J. Boyle. Ethylenediaminetetraacetic acid in manganese poisoning of rats. A preliminary study. Ind. Med. Surg. 25:1–3, 1956.

292. Kroner, R. C., and J. F. Kopp. Trace elements in six water systems of the United States. Amer. Water Works Assoc. J. 57:150–156, 1965.

293. Labanauskas, C. K. Manganese, pp. 264–285. In H. D. Chapman, Ed. Diagnostic Criteria for Plants and Soils. Riverside: University of California Division of Agricultural Sciences, 1966.

294. Labanauskas, C. K., T. W. Embleton, and W. W. Jones. Influence of soil applications of nitrogen, phosphate, potash, dolomite, and manure on the micronutrient content of avocado leaves. Proc. Amer. Soc. Hort. Sci. 71:285–291, 1958.

295. Lagerwerff, J. V. Heavy-metal contamination of soils, pp. 343–364. In N. C. Brady, Ed. Agriculture and the Quality of our Environment. A Symposium presented at the 133rd Meeting of the American Association for the Advancement of Science, Washington, D.C., 1967. Publication 85. Norwood, Mass.: The Plimpton Press, 1967.

296. Lassiter, J. W., and J. D. Morton. Effects of a low manganese diet on certain ovine characteristics. J. Anim. Sci. 27:776–779, 1968.

297. Lazrus, A. L., E. Lorange, and J. P. Lodge, Jr. Lead and other metal ions in United States precipitation. Environ. Sci. Tech. 4:55–58, 1970.

298. L-dopa in chronic manganese poisoning. Lancet 1:229, 1970. (editorial)
299. Leach, R. M., Jr. Role of manganese in the synthesis of mucopolysaccharides. Fed. Proc. 26:118-120, 1967.
300. Leach, R. M., Jr. The effect of manganese, zinc, choline and folic acid deficiencies on the composition of epiphyseal cartilage. Ph.D. thesis, Cornell University, 1960. 67 pp.
301. Leach, R. M., Jr., and A-M. Muenster. Studies on the role of manganese in bone formation. I. Effect upon the mucopolysaccharide content of chick bone. J. Nutr. 78:51-56, 1962.
302. Leach, R. M., Jr., A-M. Muenster, and E. M. Wien. Studies on the role of manganese in bone formation. II. Effect upon chondroitin sulfate synthesis in chick epiphyseal cartilage. Arch. Biochem. Biophys. 133:22-28, 1969.
303. Leeper, G. W. Manganese deficiency of cereals: plot experiments and a new hypothesis. Proc. Roy. Soc. Victoria 47:225-261, 1935.
304. Leon, A. S., H. E. Spiegel, G. Thomas, and W. B. Abrams. Pyridoxine antagonism of levodopa in parkinsonism. J.A.M.A. 218:1924-1927, 1971.
305. Leveque, L. A., and J. Beley. Contribution a l'etude de la nutrition minérale de l'arachide (*Arachis hypogaea*). Effets des toxicités borique et manganique. Agron. Trop. (Paris) 14:667-710, 1959.
306. Lindberg, O., and L. Ernster. Manganese, a co-factor of oxidative phosphorylation. Nature 173:1038-1039, 1954.
307. Lindgren, W. Mineral Deposits, p. 6. (3rd ed.) New York: McGraw-Hill Book Co., Inc., 1928.
308. Lingle, J. C., R. H. Sciaroni, B. Lear, and J. R. Wight. The effect of soil liming and fumigation on the manganese content of Brussels sprouts leaves. Proc. Amer. Soc. Hort. Sci. 78:310-318, 1961.
309. Litkins, V. A. Limits of allowable concentrations of manganese and its compounds in the atmospheric air of inhabited localities, pp. 39-51. In V. A. Ryazanov, Ed. Limits of Allowable Concentrations of Atmospheric Pollutants. Book 2. 1955. (Translated from the Russian by B. S. Levine. Office of Technical Services, U.S. Department of Commerce, Washington, D.C.)
310. Livingstone, D. A. Chemical Composition of Rivers and Lakes. pp. G1-G64. U.S. Geological Survey Professional Paper 440-G. In M. Fleischer, Ed. Data of Geochemistry. Washington, D.C.: U.S. Government Printing Office, 1963.
311. Lloyd Davies, T. A. Manganese pneumonitis. Brit. J. Ind. Med. 3:111-135, 1946.
312. Lockard, R. G. Mineral Nutrition of the Rice Plant in Malaya with Special Reference to *Penyakit Merah*. Malaya Dept. Agric. B. 108. Kuala Lumpur: The Department of Agriculture Federation of Malaya, 1959. 148 pp.
313. Lockman, R. B. Mineral composition of tobacco leaf samples. Part I. As affected by soil fertility, variety and leaf position. Soil Sci. Plant Anal. 1:95-108, 1970.
314. Lohammar, G. Wasserchemie und höhere Vegetation schwedischer Seen. Symb. Bot. Upsalienses 3:1-253, 1938-1939.
315. Löhnis, M. P. Effect of magnesium and calcium supply on the uptake of manganese by various crop plants. Plant Soil 12:339-376, 1960.
316. Löhnis, M. P. Manganese toxicity in field and market garden crops. Plant Soil 3:193-222, 1951.
317. Loneragan, J. F., J. S. Gladstones, and W. J. Simmons. Mineral elements in

References

temperate crops and pasture plants. (unpublished manuscript, University of Western Australia)

318. Loper, G. M., and D. Smith. Changes in Micronutrient Composition of the Herbage of Alfalfa, Medium Red Clover, Ladino Clover, and Brome Grass with Advance in Maturity. Wisconsin University Agricultural Experiment Station Research Report 8. Madison: University of Wisconsin Agricultural Experiment Station, 1961. 19 pp.

319. Ludwig, J. H., G. B. Morgan, and T. B. McMullen. Trends in urban air quality. Trans. Amer. Geophys. Union 51:468–475, 1970.

320. Lunt, O. R., and A. M. Kofranek. Manganese and aluminum tolerance of azalea, CV. Sweetheart Supreme, pp. 559–573. (abstract) In R. M. Samish, Ed. Recent Advances in Plant Nutrition. Vol. 2. (Abstracts of 6th International Colloquium on Plant Analysis and Fertilizer Problems, Tel-Aviv, 1970) New York: Gordon and Breach, 1970.

321. Lyon, C. B., K. C. Beeson, and G. H. Ellis. Effects of micro-nutrient deficiencies on growth and vitamin content of the tomato. Bot. Gaz. 104:495–514, 1943.

322. Lyon, M. F. Absence of otoliths in the mouse: An effect of the pallid mutant. J. Genet. 51:638–650, 1953.

323. Lyttleton, J. W. Stabilization by manganous ions of ribosomes from embryonic plant tissue. Nature 187:1026–1027, 1960.

324. Mahoney, J. P., K. Sargent, M. Greland, and W. Small. Studies on manganese. I. Determination in serum by atomic absorption spectrophotometry. Clin. Chem. 15:312–322, 1969.

325. Mahoney, J. P., and W. J. Small. Studies on manganese. III. The biological half-life of radiomanganese in man and factors which affect this half-life. J. Clin. Invest. 47:643–653, 1968.

326. Mallette, F. S. A new frontier: Air-pollution control. Inst. Mech. Eng. Proc. 168:595–615 (616–628 discussion), 1954.

327. Mantell, C. L. Manganese, pp. 271–282. In C. A. Hampel, Ed. Rare Metals Handbook. (2nd ed.) New York: Reinhold Publishing Corp., 1961.

328. Martin, G. B., D. W. Pershing, and E. E. Berkau. Effects of Fuel Additives on Air Pollutant Emissions from Distillate-Oil-Fired Furnaces. Research Triangle Park, N.C.: U.S. Environmental Protection Agency, 1971. 86 pp.

329. Mason, B. Principles of Geochemistry. (3rd ed.) New York: John Wiley and Sons, Inc., 1966. 329 pp.

330. Massey, H. F., and G. C. Johnson. Solution composition on certain coal mine spoil-bank materials as related to soil pH, pp. 73–74. In Agronomy Abstracts, 1965 Annual Meetings. Columbus: American Society of Agronomy, 1965.

331. Mathers, J. W., and R. Hill. Manganese in the nutrition and metabolism of the pullet. 2. The manganese contents of the tissues of pullets given diets of high or low manganese content. Brit. J. Nutr. 22:635–643, 1968.

332. Matrone, G., R. H. Hartman, and A. J. Clawson. Studies of a manganese-iron antagonism in the nutrition of rabbits and baby pigs. J. Nutr. 67:309–317, 1959.

333. Matteson, M. J., W. Stöber, and H. Luther. Kinetics of the oxidation of sulfur dioxide by aerosols of manganese sulfate. Ind. Eng. Chem. Fundamentals 8: 677–687, 1969.

334. Maynard, L. S., and G. C. Cotzias. The partition of manganese among organs and intracellular organelles of the rat. J. Biol. Chem. 214:489–495, 1955.
335. Maynard, L. S., and S. Fink. The influence of chelation on radiomanganese excretion in man and mouse. J. Clin. Invest. 35:831–836, 1956.
336. McCabe, L. J., and J. C. Vaughn. Trace metals content of drinking water from a large system. For presentation at Symposium on Water Quality in Distribution Systems, Division of Water, Air, and Waste Chemistry, American Chemical Society, National Meeting, Minneapolis, Minn., April 13, 1969. 21 pp.
337. McDowell, R. S. Metal carbonyl vapors: Rapid quantitative analysis by infrared spectrophotometry. Amer. Ind. Hyg. Assoc. J. 32:621–624, 1971.
338. McGannon, H. E., Ed. The Making, Shaping and Treating of Steel. (9th ed.) Pittsburgh: U.S. Steel Corp., 1971. 1420 pp.
339. McGeer, P. L., S. P. Bagchi, and E. G. McGeer. Subcellular localization of tyrosine hydroxylase in beef caudate nucleus. Life Sci. 4:1859–1867, 1965.
340. McGilvery, R. W. Biochemistry. A Functional Approach. Philadelphia: W. B. Saunders Co., 1970. 769 pp.
341. McGrath, H. Chemicals for plant disease control. Agric. Chem. 19:18–22, July 1964.
342. McKay, H. A. C. The atmospheric oxidation of sulphur dioxide in water droplets in presence of ammonia. Atmos. Environ. 5:7–14, 1971.
343. McKee, H. C., R. E. Childers, and O. Saenz, Jr. Collaborative Study of Reference Method for the Determination of Suspended Particulates in the Atmosphere (High Volume Method). San Antonio: Southwest Research Institute, 1971. 10 pp. (plus 2 appendixes)
344. McKenzie, R. E. Ability of forage plants to survive early spring flooding. Sci. Agric. 31:358–367, 1951.
345. McLennan, H. Distribution of certain compounds and enzyme systems in some areas of the nervous system, pp. 70–71. In Synaptic Transmission. Philadelphia: W. B. Saunders Company, 1963.
346. Mella, H. The experimental production of basal ganglion symptomatology in Macacus rhesus. Arch. Neurol. Psychiatr. 11:405–417, 1924.
347. Mena, I., J. Court, S. Fuenzalida, P. S. Papavasiliou, and G. C. Cotzias. Modification of chronic manganese poisoning. Treatment with L-Dopa or 5-OH tryptophane. New Eng. J. Med. 282:5–10, 1970.
348. Mena, I., K. Horiuchi, K. Burke, and G. C. Cotzias. Chronic manganese poisoning. Individual susceptibility and adsorption of iron. Neurology 19:1000–1006, 1969.
349. Mero, J. L. The Mineral Resources of the Sea. New York: Elsevier Publishing Company, 1965. 312 pp.
350. Messing, J. H. L. The effects of lime and superphosphate on manganese toxicity in steam-sterilized soil. Plant Soil 23:1–16, 1965.
351. Meyer, A. Mangan-Vergiftung, chronische, des Zentralnervensystems. Samml. Vergiftung. 1 (Sec. A-34):79–80, 1930.
352. Mildvan, A. S., M. C. Scrutton, and M. F. Utter. Pyruvate carboxylase. VII. A possible role for tightly bound manganese. J. Biol. Chem. 241:3488–3498, 1966.
353. Miller, R. C., T. B. Keith, M. A. McCarty, and W. T. S. Thorp. Manganese as a possible factor influencing the occurrence of lameness in pigs. Proc. Soc. Exp. Biol. Med. 45:50–51, 1940.

References

354. Miller, R. S., A. S. Mildvan, H-C. Chang, R. L. Easterday, H. Maruyama, and M. D. Lane. The enzymatic carboxylation of phosphoenolpyruvate. IV. The binding of manganese and substrates by phosphoenolpyruvate carboxykinase and phosphoenolpyruvate carboxylase. J. Biol. Chem. 243:6030-6040, 1968.
355. Millikan, C. R. Effect of molybdenum on the severity of toxicity symptoms in flax induced by an excess of either manganese, zinc, copper, nickel, or cobalt in the nutrient solution. J. Aust. Inst. Agric. Sci. 13:180-186, 1947.
356. Millikan, C. R. Radio-autographs of manganese in plants. Aust. J. Sci. Res. B4:28-41, 1951.
357. Misra, S. G., and P. C. Mishra. Forms of manganese as influenced by organic matter and iron oxide. Plant Soil 30:62-70, 1969.
358. Molinoff, P. B., and J. Axelrod. Biochemistry of catecholamines. Ann. Rev. Biochem. 40:465-500, 1971.
359. Molokhia, M. M., and H. Smith. Trace elements in the lung. Arch. Environ. Health 15:745-750, 1967.
360. Morgan, P. W., H. E. Joham, and J. V. Amin. Effect of manganese toxicity on the indoleacetic acid oxidase system of cotton. Plant Physiol. 41:718-724, 1966.
361. Morris, H. D. The soluble manganese content of acid soils and its relation to the growth and manganese content of sweet clover and lespedeza. Soil Sci. Soc. Amer. Proc. 13:362-371, 1948.
362. Morris, H. D., and W. H. Pierre. Minimum concentrations of manganese necessary for injury to various legumes in culture solutions. Agron. J. 41:107-112, 1949.
363. Morris, H. D., and W. H. Pierre. The effect of calcium, phosphorus, and iron on the tolerance of lespedeza to manganese toxicity in culture solutions. Soil Sci. Soc. Amer. Proc. 12:382-386, 1947.
364. Muhlrad, W. Problème des fumées émises par les fours électrométallurgiques, pp. 237-255. Chaleur Indust. No. 422. Septembre 1960.
365. Mulder, E. G., and F. C. Gerretsen. Soil manganese in relation to plant growth, pp. 221-277. In A. G. Norman, Ed. Advances in Agronomy. Vol. IV. New York: Academic Press, 1952.
366. Munns, D. N., C. M. Johnson, and L. Jacobson. Uptake and distribution of manganese in oat plants. I. Varietal variation. Plant Soil 19:115-126, 1963.
367. Munns, D. N., C. M. Johnson, and L. Jacobson. Uptake and distribution of manganese in oat plants. III. An analysis of biotic and environmental effects. Plant Soil 19:285-295, 1963.
368. Munro, I. B. Infectious and non-infectious herd infertility in East Anglia. Vet. Rec. 69:125-129, 1957.
369. Nagatsu, T., M. Levitt, and S. Udenfriend. Tyrosine hydroxylase. The initial step in norepinephrine biosynthesis. J. Biol. Chem. 239:2910-2917, 1964.
370. Nason, A., and W. D. McElroy. Modes of action of the essential mineral elements, pp. 451-536. In F. C. Steward, Ed. Plant Physiology. A Treatise. Vol. III. Inorganic Nutrition of Plants. New York: Academic Press, 1963.
371. National Academy of Sciences. Recommended Dietary Allowances. (7th rev. ed.) A Report of the Food and Nutrition Board, National Research Council. NAS Publication 1694. Washington, D.C.: National Academy of Sciences, 1968. 101 pp.

372. Nazif, M. Manganese as an industrial poisoning. J. Egypt. Public Health Assoc. 10:1-20, 1936.
373. Neenan, M. The effects of soil acidity on the growth of cereals with particular reference to the differential reaction of varieties thereto. Plant Soil 12:324-338, 1960.
374. Neff, N. H., R. E. Barrett, and E. Costa. Selective depletion of caudate nucleus dopamine and serotonin during chronic manganese dioxide administration to squirrel monkeys. Experientia 25:1140-1141, 1969.
375. Nelson, W. L., and S. A. Barber. Nutrient deficiencies in legumes for grain and forage, pp. 143-180. In H. B. Sprague, Ed. Hunger Signs in Crops: A Symposium. (3rd ed.) New York: David McKay Co., 1964.
376. Nicholas, D. J. D. Detection of manganese deficiency in plants by tissue tests using tetramethyldiaminodiphenylmethane. Nature 157:696, 1946.
377. Nifong, G. D., E. A. Boettner, and J. W. Winchester. Particle size distributions of trace elements in pollution aerosols. Amer. Ind. Hyg. Assoc. J. (In press)
378. Norris, L. C., and C. D. Caskey. A chronic congenital ataxia and osteodystrophy in chicks due to manganese deficiency. J. Nutr. 17(Suppl.):16-17, 1939. (abstract)
379. North, B. B., J. M. Leichsenring, and L. M. Norris. Manganese metabolism in college women. J. Nutr. 72:217-223, 1960.
380. Nyborg, M. Sensitivity to manganese deficiency of different cultivars of wheat, oats, and barley. Can. J. Plant Sci. 50:198-200, 1970.
381. O'Dell, B. L., and B. J. Campbell. Trace elements: Metabolism and metabolic function, pp. 179-266. In M. Florkin and E. H. Stotz, Eds. Comprehensive Biochemistry. Vol. 21. Metabolism of Vitamins and Trace Elements. New York: American Elsevier Publishing Co., Inc., 1971.
382. Ohle, W. Chemische und physikalische Untersuchungen norddeutscher Seen. Arch. Hydrobiol. 26:386-464, 584-658, 1934.
383. Olsen, C. Absorption of manganese by plants. II. Toxicity of manganese to various plant species. C. R. Trav. Lab. Carlsberg 21:Ser. Chim. 129-145, 1936.
384. Oltramare, M., M. Tchicaloff, P. Desbaumes, and G. Hermann. Intoxication chronique au manganèse chez deux soudeurs à l'arc. Arch. Gewerbepathol. Gewerbehyg. 21:124-140, 1965.
385. Orten, J. M., and O. W. Neuhaus. Biochemistry. (8th ed.) St. Louis: C. V. Mosby Co., 1970. 925 pp.
386. Ouellette, G. J., and L. Dessureaux. Chemical composition of alfalfa as related to degree of tolerance to manganese and aluminum. Can. J. Plant Sci. 38:206-214, 1958.
387. Ouellette, G. J. Toxicité du manganèse dans les sols fortement acides. Agriculture (Quebec) 7:319-322, 1950.
388. Ouellette, G. J., and H. Genereaux. Influence de l'intoxication manganique sur six variétès de pomme de terre. Can. J. Soil Sci. 45:24-32, 1965.
389. Pancheri, G. Industrial atmospheric pollution in Italy, pp. 252-263. In F. S. Mallette, Ed. Problems and Control of Air Pollution. Proceedings of the First International Congress on Air Pollution, New York City, March 1-2, 1955, under the sponsorship of the Committee on Air-Pollution Controls of The American Society of Mechanical Engineers. New York: Reinhold Publishing Corporation, 1955.
390. Papavasiliou, P. S., S. T. Miller, and G. C. Cotzias. Functional interactions be-

tween biogenic amines, 3',5'-cyclic AMP and manganese. Nature 220:74-75, 1968.
391. Papavasiliou, P. S., S. T. Miller, and G. C. Cotzias. Role of liver in regulating distribution and excretion of manganese. Amer. J. Physiol. 211:211-216, 1966.
392. Parbery, N. H. The excessive uptake of manganese by beans, showing scald and magnesium deficiency. Its regulation by liming. Agric. Gaz. N. S. Wales 54:14-17, 1943.
393. Park, R. B., and N. G. Pon. Chemical composition and the substructure of lamellae isolated from *Spinacea oleracae* chloroplasts. J. Mol. Biol. 6:105-114, 1963.
394. Parker, M. B., H. B. Harris, H. D. Morris, and H. F. Perkins. Manganese toxicity of soybeans as related to soil and fertility treatments. Agron. J. 61:515-521, 1969.
395. Peñalver, R. Diagnosis and treatment of manganese intoxication. Report of a case. A.M.A. Arch. Ind. Health 16:64-66, 1957.
396. Peñalver, R. Manganese poisoning: The 1954 Ramazzini oration. Ind. Med. Surg. 24:1-7, 1955.
397. Pentschew, A., F. F. Ebner, and R. M. Kovatch. Experimental manganese encephalopathy in monkeys. A preliminary report. J. Neuropath. Exp. Neurol. 22:488-499, 1963.
398. Pentschew, A., A. Riopelle, R. Kovatch, and P. Lampert. Regrouping of the extrapyramidal diseases as suggested by manganese encephalopathy. Implications on treatment of Parkinson's disease with manganese. (The Second Symposium on Parkinson's Disease, Washington, D.C., Nov. 18-20, 1963.) J. Neurosurg. 24 (Suppl., Part II):255, 1966.
399. Perkins, E. C., and F. Noviello. Bacterial Leaching of Manganese Ores. U.S. Bureau of Mines Report of Investigations 6102. Washington, D.C.: U.S. Department of the Interior, 1962. 11 pp.
400. Person, R. A. Control of emissions from ferroalloy furnace processing, pp. 81-92. In Proceedings of 27th Electric Furnace Conference; Iron and Steel Division of the Metallurgical Society of the American Institute of Mining, Metallurgical and Petroleum Engineers, Detroit, December 10-12, 1969. New York: American Institute of Mining, Metallurgical and Petroleum Engineers, 1970.
401. Petering, H. G., D. W. Yeager, and S. O. Witherup. Trace metal content of hair. I. Zinc and copper content of human hair in relation to age and sex. Arch. Environ. Health 23:202-207, 1971.
402. Peters, D. A. V., P. L. McGeer, and E. G. McGeer. The distribution of tryptophan hydroxylase in cat brain. J. Neurochem. 15:1431-1435, 1968.
403. Peterson, W. H., and J. T. Skinner. Distribution of manganese in foods. J. Nutr. 4:419-426, 1931.
404. Pinkerton, C., D. I. Hammer, T. Hinners, V. Hasselblad, J. Kent, J. V. Lagerwerff, and E. F. Ferrand. Trace metals in urban soils and house dust. Presented at the 100th Annual Meeting of the American Public Health Association, 12-16 November 1972, Atlantic City, N.J.
405. Pletscher, A., G. Bartholini, and R. Tissot. Metabolic fate of L[^{14}C]DOPA in cerebrospinal fluid and blood plasma of humans. Brain Res. 4:106-109, 1967.
406. Plumlee, M. P., D. M. Thrasher, W. M. Beeson, F. N. Andrews, and H. E. Parker. The effects of a manganese deficiency on the growth and develop-

ment of swine. J. Anim. Sci. 13:996, 1954. (abstract) (Abstracts of papers to be presented at 46th Annual Meeting of American Society of Animal Production)
407. Plumlee, M. P., D. M. Thrasher, W. M. Beeson, F. N. Andrews, and H. E. Parker. The effects of a manganese deficiency upon the growth, development, and reproduction of swine. J. Anim. Sci. 15:352–367, 1956.
408. Possingham, J. V., M. Vesk, and F. V. Mercer. The fine structure of leaf cells of manganese-deficient spinach. J. Ultrastruct. Res. 11:68–83, 1964.
409. Price, N. O., W. N. Linkous, and R. W. Engel. Minor element content of forage plants and soils. J. Agric. Food Chem. 3:226–229, 1955.
410. Pringle, B. H., D. E. Hissong, E. L. Katz, and S. T. Mulawka. Trace metal accumulation by estuarine mollusks. J. Sanit. Eng. Div. Proc. Amer. Soc. Civil Engrs. 94:455–475, 1968.
411. Prosperi, G., and C. Barsi. Sulla patologia dei lavoratori del manganese. Rass. Med. Ind. 23:316–320, 1954.
412. Pullman, A., and B. Pullman. The band structure of melanins. Biochim. Biophys. Acta 54:384–385, 1961.
413. Quality goals for potable water–statement of policy. J. Amer. Water Works Assoc. 60:1317–1322, 1968.
414. Rees, W. J., and G. H. Sidrak. Inter-relationship of aluminum and manganese toxicities towards plants. Plant Soil 14:101–117, 1961.
415. Reid, W. S. The effect of several factors on the response of barley (*Hordeium vulgare*, L.) to excess soil manganese in an acid terrace soil. Ph.D. thesis, Mississippi University, 1965. Diss. Abstr. 26(9):4947.
416. Reuther, W., P. F. Smith, and A. W. Specht. Accumulation of the major bases and heavy metals in Florida citrus soils in relation to phosphate fertilization. Soil Sci. 73:375–381, 1952.
417. Reuther, W., P. F. Smith, and A. W. Specht. A comparison of the mineral composition of Valencia orange leaves from the major producing areas of the United States. Proc. Florida State Hort. Soc. 62:38–45, 1949.
418. Riddervold. pp. 2546–2547. In Discussion of a paper by D. Elstad. Factory smoke containing manganese as contributing cause in pneumonia epidemics in an industrial district. Nord. Med. 3:2527–2533, 1939. (in Norwegian)
419. Riopelle, A. J. Manganese effects in chimpanzees, pp. 47–52. In G. E. Crane and R. Gardner, Jr., Eds. Psychotropic Drugs and Dysfunctions of the Basal Ganglia: A Multidisciplinary Workshop. Public Health Service Publication 1938. Washington, D.C.: U.S. Department of Health, Education, and Welfare, 1969.
420. Rippel, A. The iron chlorosis in green plants caused by manganese. Biochem. Z. 140:315–323, 1923.
421. Ritter, J. Ch. Manganisme dans les mines d'Idikel–Tafraout. Maroc Med. 37: 455–462, 1958.
422. Rjazanov, V. A. Criteria and methods for establishing maximum permissible concentrations of air pollution. Bull. WHO 32:389–398, 1965.
423. Robinson, D. B., and W. A. Hodgson. Note on the effect of some amino acids on manganese toxicity in potato. Can. J. Plant Sci. 41:436–437, 1961.
424. Robinson, N. W., S. L. Hansard, D. M. Johns, and G. L. Robertson. Excess dietary manganese and feed lot performance of beef cattle. J. Anim. Sci. 19:

1290, 1960. (abstract) (Abstracts of papers for presentation at 52nd Annual Meeting of the American Society of Animal Production)
425. Robson, A. D., and J. F. Loneragan. Sensitivity of annual *Medicago* species to manganese toxicity as affected by calcium and pH. Aust. J. Agric. Res. 21: 223–232, 1970.
426. Rodier, J. Le manganisme au Maroc. Maroc Med. 37:429–454, 1958.
427. Rodier, J. Manganese poisoning in Moroccan miners. Brit. J. Ind. Med. 12:21–35, 1955.
428. Rodier, J., R. Mallet, and L. Rodi. Étude de l'action détoxicante de l'ethylénediaminetetraacétate de calcium dans l'intoxication expérimentale par le manganese. Arch. Mal. Prof. 15:210–223, 1954.
429. Rodier, J., and M. Rodier. Le manganisme dans les mines marocaines. Bull. Inst. Hyg. Maroc 9 (N.S.):3–98, 1949.
430. Rojas, M. A., I. A. Dyer, and W. A. Cassatt. Manganese deficiency in the bovine. J. Anim. Sci. 24:664–667, 1965.
431. Roldan V., U. Manganismo profesional. Observaciones clinicas, pp. 100–104. In Tercer Congreso Americano de Medicina del Trabajo, Vol. I., Caracas, 1955. Buenos Aires: Union Americana de Medicine del Trabajo, 1956.
432. Romney, E. M., and S. J. Toth. Plant and soil studies with radioactive manganese. Soil Sci. 77:107–117, 1954.
433. Rosenstock, H. A., D. G. Simons, and J. S. Meyer. Chronic manganism. Neurologic and laboratory studies during treatment with levodopa. J.A.M.A. 217: 1354–1358, 1971.
434. Runnels, R. T., and J. A. Schleicher. Chemical composition of eastern Kansas limestones, pp. 81–103. In State Geological Survey of Kansas Bulletin 119, Part 3. Lawrence, Kans.: University of Kansas Publications, 1956.
435. Ryzhkova, M. N., G. N. Cherepanova, and R. L. Blekh. Early diagnosis of chronic manganese intoxication. Tr. Akad. Meditsink. Nauk SSSR (Moskva) 31:35–43, 1954. (in Russian)
436. Sanchez, C., and E. J. Kamprath. Effect of liming and organic matter content on the availability of native and applied manganese. Soil Sci. Soc. Amer. Proc. 23:302–304, 1959.
437. Sax, N. I. Dangerous Properties of Industrial Materials. (2nd ed.) New York: Reinhold Publishing Corp., 1963. 1343 pp.
438. Scander, A., and H. A. Sallam. A report on eleven cases of chronic manganese poisoning. J. Egypt. Med. Assoc. 19:57–62, 1936.
439. Schlockow. Ueber ein eigenartiges Rückenmarksleiden der Zinkhüttenarbeiter. Dtsch. Med. Wochenschr. 5:208–210, 221–222, 1879.
440. Schmehl, W. R., and R. P. Humbert. Nutrient deficiencies in sugar crops, pp. 415–450. In H. B. Sprague, Ed. Hunger Signs in Crops: A Symposium. (3rd ed.) New York: David McKay Co., 1964.
441. Schopper, W. Über Lungenentzündungen bei Brauneisenstein-Bergarbeitern. Arch. Hyg. Bakt. 104:175–183, 1930.
442. Schroeder, H. A. Manganese. Air Quality Monograph 70-17. Washington, D.C.: American Petroleum Institute, 1970. 34 pp.
443. Schroeder, H. A., J. J. Balassa, and I. H. Tipton. Essential trace metals in man: manganese. A study in homeostasis. J. Chron. Dis. 19:545–571, 1966.
444. Schuler, P., H. Oyanguren, V. Maturana, A. Valenzuela, E. Cruz, V. Plaza,

E. Schmidt, and R. Haddad. Manganese poisoning. Environmental and medical study at a Chilean mine. Ind. Med. Surg. 26:167–173, 1957.
445. Scrutton, M. C., M. F. Utter, and A. S. Mildvan. Pyruvate carboxylase. VI. The presence of tightly bound manganese. J. Biol. Chem. 241:3480–3487, 1966.
446. Shacklette, H. T., J. C. Hamilton, J. G. Boerngen, and J. M. Bowles. Elemental Composition of Surficial Materials in the Conterminous United States. U.S. Geological Survey Paper 574-D. Washington, D.C.: U.S. Government Printing Office, 1971. 71 pp.
447. Sheldon, J. H. Some considerations on influence of copper and manganese on the therapeutic activity of iron. Brit. Med. J. 2:869–872, 1932.
448. Shelton, J. E., and D. C. Zeiger. Distribution of manganese^{-54} in 'Delicious' apple trees in relation to the occurrence of internal bark necrosis (IBN). J. Amer. Soc. Hort. Sci. 95:758–762, 1970.
449. Sherman, G. D., and P. M. Harmer. The manganous-manganic equilibrium of soils. Soil Sci. Soc. Amer. Proc. 7:398–405, 1942.
450. Sherman, G. D., J. S. McHargue, and W. S. Hodgkiss. Determination of active manganese in soil. Soil Sci. 54:253–257, 1942.
451. Shils, M. E., and E. V. McCollum. Further studies on the symptoms of manganese deficiency in the rat and mouse. J. Nutr. 26:1–19, 1943.
452. Shrader, R. E., and G. J. Everson. Anomalous development of otoliths associated with postural defects in manganese-deficient guinea pigs. J. Nutr. 91: 453–460, 1967.
453. Singh, A., A. Kambal, and S. S. Singh. The iron–manganese relationships in plant growth. Indian J. Agron. 6:298–303, 1962.
454. Singh, M., and A. N. Pathak. Effect of heating and steam sterilization on soil manganese. Plant Soil 33:244–248, 1970.
455. Single, W. V., and I. F. Bird. The mobility of manganese in the wheat plant. II. Redistribution in relation to concentration and chemical state. Ann. Bot. N.S. 22:489–502, 1958.
456. Skougstad, M. W. Minor elements in water, pp. 43–55. In H. L. Cannon and H. C. Hopps, Eds. Environmental Geochemistry in Health and Disease. American Association for Advancement of Science Symposium, Dallas, Texas, December 1968. Geological Society of America, Inc., Memoir 123. Boulder, Col.: The Geological Society of America, Inc., 1971.
457. Slowey, J. I. Studies on the distribution of copper, manganese, and zinc in the ocean using neutron activation analysis, pp. 1–105. In The Chemistry and Analysis of Trace Metals in Sea Water. Final Report, AEC Contract AT(40-1)-2799. College Station, Texas: Texas A & M University, 1966.
458. Smith, M. W., and J. B. Storey. The analysis of pecan leaves by atomic absorption spectroscopy. Soil Sci. Plant Anal. 2:249–258, 1971.
459. Smith, S. E., M. Medlicott, and G. H. Ellis. Manganese deficiency in the rabbit. Arch. Biochem. 4:281–289, 1944.
460. Smith, W. S. Atmospheric Emissions from Fuel Oil Combustion—An Inventory Guide, pp. 32–37. U.S. Public Health Service Publication 999-AP-2. Division of Air Pollution, Public Health Service, U.S. Department of Health, Education, and Welfare. Cincinnati: U.S. Department of Health, Education, and Welfare, 1962.

References

461. Somers, I. I., and J. W. Shive. The iron–manganese relation in plant metabolism. Plant Physiol. 17:582–602, 1942.
462. Sourkes, T. L. Dopa decarboxylase: Substrates, coenzyme, inhibitors. (in Second Symposium on Catecholamines, Milano, 1965) Pharmacol. Rev. 18: 53–60, 1966.
463. Sparr, M. C. Micronutrient needs–which, where, on what–in the United States. Soil Sci. Plant Anal. 1:241–262, 1970.
464. Starodubova, T. F. Manganese balance in adolescents-students of boarding schools during the academic year and summer vacation in a pioneer's camp. Vopr. Pitan. 27(4):36–40, 1968. (in Russian; summary in English)
465. Stöcker, W. Ein Fall von fortschreitender Lenticulardegeneration. Z. Gesamte Neurol. Psychiat. 15:251–272, 1913.
466. Stokinger, H. E. The metals (excluding lead), pp. 987–1217. In F. A. Patty, Ed. Industrial Hygiene and Toxicology. (2nd rev. ed.) D. W. Fassett and D. D. Irish, Eds. Vol. II. Toxicology, pp. 1087–1089. New York: Interscience Publishers, 1962.
467. Stonier, T., F. Rodriguez-Tormes, and Y. Yoneda. Studies on auxin protectors. IV. The effect of manganese on auxin protector-I of the Japanese morning glory. Plant Physiol. 43:69–72, 1968.
468. Sully, A. H. Manganese. (Metallurgy of the Rarer Metals, No. 3.) New York: Academic Press, Inc., 1955. 305 pp.
469. Sutcliffe, J. F. Mineral Salts Absorption in Plants. (International Series of Monographs on Pure and Applied Biology: Plant Physiology Division. Vol. 1.) New York: Pergamon Press, Inc., 1962. 194 pp.
470. Sutherland, E. W., and G. A. Robison. Metabolic effects of catecholamines. A. The role of cyclic-$3',5'$-AMP in responses to catecholamines and other hormones. Pharmacol. Rev. 18:145–161, 1966.
471. Sutton, C. D., and E. G. Hallsworth. Studies on the nutrition of forage legumes. I. The toxicity of low pH and high manganese supply to lucerne, as affected by climatic factors and calcium supply. Plant Soil 9:305–317, 1958.
472. Suzuki, Y., K. Nishiyama, M. Doi, T. Hirose, H. Shibata. Studies on chronic manganese poisoning. Tokushima J. Exp. Med. 7:124–132, 1960.
473. Sverdrup, H. U., M. W. Johnson, and R. H. Fleming. The Oceans–Their Physics, Chemistry, and General Biology, pp. 176–177. 2nd printing, New York: Prentice-Hall, Inc., 1946.
474. Swaine, D. J. The Trace-element Content of Soils. Commonwealth Bureau of Soil Science Technical Communication 48. York, England: Herald Printing Works, 1955. 157 pp.
475. Swaine, D. J., and R. L. Mitchell. Trace-element distribution in soil profiles. J. Soil Science 11:347–368, 1960.
476. Tabor, E. C., and W. V. Warren. Distribution of certain metals in the atmosphere of some American cities. A.M.A. Arch. Ind. Health 17:145–151, 1958.
477. Tal, E., and K. Guggenheim. Effect of manganese on calcification of bone. Biochem. J. 95:94–97, 1965.
478. Tanaka, A., and S. A. Navasero. Manganese content of the rice plant under water culture conditions. Soil Sci. Plant. Nutr. 12:21–26, 1966.
479. Tanaka, S., and J. Lieben. Manganese poisoning and exposure in Pennsylvania. Arch. Environ. Health 19:674–684, 1969.

480. Taras, M. J., A. E. Greenberg, R. D. Hoak, and M. C. Rand, Joint Editorial Board. Standard Methods for the Examination of Water and Wastewater. (13th ed.) Prepared and published jointly by the American Public Health Association, the American Water Works Association and the Water Pollution Control Federation. Washington, D.C.: American Public Health Association, 1971. 874 pp.
481. Tauber, F. W., and A. C. Krause. The role of iron, copper, zinc and manganese in the metabolism of the ocular tissues, with special reference to the lens. Amer. J. Ophthalmol. 26:260–266, 1943.
482. Taylor, D. M. The manganese nutrition of cotton. Ph.D. thesis, Texas A. and M. University, 1965. Diss. Abstr. 26(1):50.
483. Taylor, W. G. Smoke Elimination in Gas Turbines Burning Distillate Oil. Presented at ASME–AIEE Joint Power Generation Conference, Detroit, Mich., Sept. 24–28, 1967. New York: American Society of Mechanical Engineers. 9 pp.
484. Tentative method of analysis for manganese content of atmospheric particulate matter. Health Lab. Sci. 7:146–148, 1970.
485. Tepper, L. B. Seven-City Study of Air and Population Lead Levels. An Interim Report. Department of Environmental Health, College of Medicine, University of Cincinnati, 1971, 11 pp.
486. Thiers, R. E., and B. L. Vallee. Distribution of metals in subcellular fractions of rat liver. J. Biol. Chem. 226:911–920, 1957.
487. Thompson, R. J., G. B. Morgan, and L. J. Purdue. Analysis of selected elements in atmospheric particulate matter by atomic absorption. Atomic Absorption Newslett. 9:53–57, 1970.
488. Tietz, N. W., E. F. Hirsch, and B. Neyman. Spectrographic study of trace elements in cancerous and noncancerous human tissues. J.A.M.A. 165:2187–2192, 1957.
489. Tiffin, L. O. Translocation of micronutrients in plants, pp. 199–229. In J. J. Mortvedt, P. M. Giordano, and W. L. Lindsay, Eds. Micronutrients in Agriculture. Proceedings of a symposium held at Muscle Shoals, Ala., April 20–22, 1971, and cosponsored by the Tennessee Valley Authority and the Soil Science Society of America. Madison, Wisc.: Soil Science Society of America, Inc., 1972.
490. Timonin, M. I., and G. R. Giles. Effect of different soil treatments on microbial activity and availability of manganese in manganese deficient soil. J. Soil Sci. 3:145–155, 1952.
491. Tipton, I. H., and M. J. Cook. Trace elements in human tissue. II. Adult subjects from the United States. Health Phys. 9:103–145, 1963.
492. Tipton, I. H., M. J. Cook, R. L. Steiner, C. A. Boye, H. M. Perry, Jr., and H. A. Schroeder. Trace elements in human tissue. I. Methods. Health Phys. 9:89–101, 1963.
493. Tissot, R., G. Bartholini, and A. Pletscher. Drug-induced changes of extracerebral dopa metabolism in man. Arch. Neurol. 20:187–190, 1969.
494. Toth, S. J., and E. M. Romney. Manganese studies with some New Jersey soils. Soil Sci. 78:295–303, 1954.
495. Tsai, H. C., and G. J. Everson. Effect of manganese deficiency on the acid mucopolysaccharides in cartilage of guinea pigs. J. Nutr. 91:447–452, 1967.
496. Turekian, K. K. The oceans, streams, and atmosphere, pp. 297–323. In K. H.

References

Wedepohl, Ed. Handbook of Geochemistry. Vol. I. New York: Springer-Verlag, 1969.
497. Turekian, K. K., and K. H. Wedepohl. Distribution of the elements in some major units of the earth's crust. Geol. Soc. Amer. Bull. 72:175-192, 1961.
498. Tyson, A. G. Manganese deficiency in subterranean clover (*Trifolium subterraneum* L.). Aust. J. Agric. Res. 5:608-613, 1954.
499. Udenfriend, S. Tyrosine hydroxylase. Pharmacol. Rev. 18:43-51, 1966.
500. Underwood, E. J. Trace Elements in Human and Animal Nutrition, p. 180. (3rd ed.) New York: Academic Press, 1971.
501. U.S. Department of Health, Education, and Welfare, National Air Pollution Control Administration. Air Quality Data from the National Air Sampling Networks and Contributing State and Local Networks. 1966 Edition. NAPCA Publication APTD 68-9. Durham, N.C.: U.S. Department of Health, Education, and Welfare, 1968. 157 pp.
502. U.S. Department of Health, Education, and Welfare, Public Health Service, Bureau of State Services, Division of Sanitary Engineering Services, Robert A. Taft Sanitary Engineering Center, Cincinnati, Ohio. Air Pollution Measurements of the National Air Sampling Network. Analysis of Suspended Particulate Samples Collected 1953-1957. PHS Publication 637. Washington, D.C.: U.S. Government Printing Office, 1958. 259 pp.
503. U.S. Department of Health, Education, and Welfare, Public Health Service, Consumer Protection and Environmental Health Service, Environmental Control Administration. Public Health Service Drinking Water Standards. Revised 1962. Rockville, Md.: U.S. Department of Health, Education, and Welfare, 1972. 61 pp.
504. U.S. Department of Health, Education, and Welfare, Public Health Service, Division of Air Pollution, Robert A. Taft Sanitary Engineering Center, Cincinnati, Ohio. Air Pollution Measurements of the National Air Sampling Network. Analyses of Suspended Particulates, 1957-1961. PHS Publication 978. Washington, D.C.: U.S. Government Printing Office, 1962. 217 pp.
505. U.S. Department of Health, Education, and Welfare, Public Health Service, Division of Air Pollution, Robert A. Taft Sanitary Engineering Center. Air Pollution Measurements of the National Air Sampling Network. Analyses of Suspended Particulates, 1963. Cincinnati: U.S. Department of Health, Education, and Welfare, 1965. 125 pp.
506. U.S. Department of Health, Education, and Welfare, Public Health Service, Division of Air Pollution, Robert A. Taft Sanitary Engineering Center. Air Quality Data, National Air Sampling Network, 1962. Cincinnati: U.S. Department of Health, Education, and Welfare. 50 pp.
507. U.S. Department of Health, Education, and Welfare, Public Health Service, Division of Air Pollution, Robert A. Taft Sanitary Engineering Center, Air Quality Data from the National Air Sampling Networks and Contributing State and Local Networks, 1964-1965. Cincinnati: U.S. Department of Health, Education, and Welfare, 1966. 106 pp.
508. U.S. Department of Health, Education, and Welfare, Public Health Service, Environmental Health Service, Bureau of Water Hygiene. Community Water Supply Study. Analysis of National Survey Findings. Washington, D.C.: U.S. Department of Health, Education, and Welfare, 1970. 123 pp.

509. U.S. Department of the Interior. Federal Water Pollution Control Administration. Report of the Committee on Water Quality Criteria. Washington, D.C.: U.S. Government Printing Office, 1968. 234 pp.
510. U.S. Environmental Protection Agency, Division of Atmospheric Surveillance. Air Quality Data for 1967 from the National Air Surveillance Networks and Contributing State and Local Networks. Revised 1971. Office of Air Programs. APTD Publication 0741. Research Triangle Park, N.C.: U.S. Environmental Protection Agency, 1971. 184 pp.
511. Vacek, A., and A. Schertler. Waste-gas cleaning systems at oxygen steel plants, pp. 82-89. In Iron and Steel Industry Report 61. Air and Water Pollution in the Iron and Steel Industry. Proceedings of the Air Pollution Meeting, 25-26 September and the Water Pollution Meeting, 11-12 December, 1957, London. London: The Iron and Steel Institute, 1958.
512. Vallee, B. L., and J. E. Coleman. Metal coordination and enzyme action, pp. 165-235. In M. Florkin and E. H. Stotz, Eds. Comprehensive Biochemistry. Vol. 12. Enzymes: General Considerations. New York: Elsevier Publishing Company, 1964.
513. Van Beukering, J. A. The occurrence of pneumonia among miners in an iron mine and a manganese ore mine in South Africa. Ned. Tijdschr. Geneeskd. 110:473-474, 1966. (in Dutch)
514. Van Bogaert, L., and M. J. Dallemagne. Approches expérimentales des troubles nerveux du manganisme. Monatsschr. Psychiat. Neurol. 111:60-89, 1945-46.
515. Van Koetsveld, E. E. The manganese and copper contents of hair as an indication of the feeding condition of cattle regarding manganese and copper. Tijdschr. Diergeneesk. 83:229-236, 1958.
516. Vinogradov, A. P. The Elementary Chemical Composition of Marine Organisms. Sears Foundation for Marine Research Memoir II. Yale University, New Haven, Connecticut. Copenhagen, Denmark: Bianco Luno's Printing, 1953. 647 pp.
517. Vinogradov, A. P. The Geochemistry of Rare and Dispersed Chemical Elements in Soils. (2nd ed.) London: Chapman and Hall, 1959. 209 pp.
518. Vlamis, J., and D. E. Williams. Ion competition in manganese uptake by barley plants. Plant Physiol. 37:650-655, 1962.
519. Vlamis, J., and D. E. Williams. Iron and manganese relations in rice and barley. Plant Soil 20:221-231, 1964.
520. Vlamis, J., and D. E. Williams. Manganese and silicon interaction in the Gramineae. Plant Soil 27:131-140, 1967.
521. Von Oettingen, W. F. Manganese: Its distribution, pharmacology and health hazards. Physiol. Rev. 15:175-201, 1935.
522. Vose, P. B. The translocation and redistribution of manganese in Avena. J. Exp. Bot. 14:448-457, 1963.
523. Vose, P. B., and D. J. Griffiths. Manganese and magnesium in the grey speck syndrome of oats. Nature 191:299-300, 1961.
524. Voskian, H., F. Rousselet, and M. L. Girard. Possibilités de dosage du manganèse dans les milieux biologiques, pp. 203-234. In La Photométrie d'Absorption Atomique dans la Flamme; Colloque d'Information Scientifique, Gembloux, le 28 mai 1968. Gembloux, Glegium: Centre de Récherches Angronomiques de l'Etat, 1968.

525. Voss, H. Über das Vorkommen von beruflichen Manganvergiftungen in der Stahlindustrie (anlässlich eines Falles von Manganismus bei einem Ferromanganmüller). Arch. Gewerbepath. Gewerbehyg. 9:453–463, 1939.
526. Wachtel, L. W., C. A. Elvehjem, and E. B. Hart. Studies on physiology of manganese in the rat. Amer. J. Physiol. 140:72–82, 1943.
527. Walford, L. A. Living Resources of the Sea: Opportunities for Research and Expansion. New York: The Ronald Press Co., 1958. 321 pp.
528. Wallace, A. Regulation of the Micronutrient Status of Plants by Chelating Agents and Other Factors, pp. 18, 87, and 257. Los Angeles: Arthur Wallace, Publisher, 1971.
529. Wallace, A., and R. T. Mueller. Effect of chelating agents on the availability of ^{54}Mn following its addition as carrier-free ^{54}Mn to three different soils. Soil Sci. Soc. Amer. Proc. 32:828–830, 1968.
530. Warren, H. V., R. E. Delavault, and K. W. Fletcher. Metal pollution. A growing problem in industrial and urban areas. Can. Mining Met. Bull. 64:34–35, 1971.
531. Wassermann, M., and G. Mihail. Indicateurs significatifs pour le dépistage précoce du manganisme chez les mineurs des mines de manganèse. Acta Med. Leg. Soc. 17:61–89, 1964.
532. Wassermann, M., V. Voiculescu, B. Polingher, A. Solomonovici, Gh. Pendefunda, St. Rucinschi, I. Kleinstein, A. Ionescu, H. Feller, and O. Antonovici. Contribution to the understanding and prevention of manganism in manganese mines in the People's Republic of Rumania. Academia Republicii Populare Romine. Filiala Iasi. Studii si Ceretari Stiintifice. Seria 2, Stiinte Biologice, Medicale si Agricole. 5(3–4):213–224, 1954. (in Rumanian)
533. Watson, G. A. The effect of soil pH and manganese toxicity upon the growth and mineral composition of the hop plant. J. Hort. Sci. 35:136–145, 1960.
534. Wende, C. V., and M. T. Spoerlein. Oxidation of dopamine to melanin by an enzyme of rat brain. Life Sci. 2:386–392, 1963.
535. Werner, A., and M. Anke. Der Spurenelementgehalt der Rinderhaare als Hilfsmittel zur Erkennung von Mangelerscheinungen. Arch. Tierernaehr. 10:142–153, 1960.
536. Westermann, D. T., T. L. Jackson, and D. P. Moore. Effect of potassium salts on extractable soil manganese. Soil Sci. Soc. Amer. Proc. 35:43–46, 1971.
537. White, R. P. Effects of lime upon soil and plant manganese levels in an acid soil. Soil Sci. Soc. Amer. Proc. 34:625–629, 1970.
538. Whitlock, C. M., Jr., S. J. Amuso, and J. B. Bittenbender. Chronic neurological disease in two manganese steel workers. Amer. Ind. Hyg. Assoc. J. 27:454–459, 1966.
539. Wiebe, A. H. The manganese content of the Mississippi River water at Fairport, Iowa. Science 71:248, 1930.
540. Wiebe, H.-J. Manganvergiftungen bei Kopfsalat als Folge der Bodendämpfung. Gemuese 5:226–227, 1969.
541. Wiebe, H.-J., and P. Wetzold. Nebenwirkungen der Bodensterilisation mit Dampf auf die Entwicklung von Kopfsalat (*Lactuca sativa* var. *capitata* L.). Z. Pflanzenkrank. (Pflanzenpath.) Pflanzenschutz 76:340–348, 1969.
542. Wiese, A. C., B. C. Johnson, C. A. Elvehjem, E. B. Hart, and J. G. Halpin. A study of blood and bone phosphatase in chick perosis. J. Biol. Chem. 127:411–420, 1939.

543. Williams, D. E., and J. Vlamis. Manganese and boron toxicities in standard culture solutions. Soil Sci. Soc. Amer. Proc. 21:205–209, 1957.
544. Williams, D. E., and J. Vlamis. The effect of silicon on yield and manganese-54 uptake and distribution in the leaves of barley plants grown in culture solutions. Plant Physiol. 32:404–409, 1957.
545. Williams, P. C. Nickel, iron, and manganese in the metabolism of the oat plant. Nature 214:628, 1927.
546. Williams, R. J., and E. M. Lansford, Jr., Eds. The Encyclopedia of Biochemistry. New York: Reinhold Publishing Corporation, 1967. 876 pp.
547. Williams, R. J. P. Coordination, chelation, and catalysis, pp. 391–441. In P. D. Boyer, H. Lardy, and K. Myrbäck, Eds. The Enzymes. Vol. 1. Kinetics, Thermodynamics, Mechanism, Basic Properties. 2nd ed. (rev.) New York: Academic Press, 1959.
548. Williams, R. J. P. Heavy metals in biological systems. Endeavour 26:96–100, 1967.
549. Wilson, J. G. Bovine functional infertility in Devon and Cornwall: Response to manganese therapy. Vet. Rec. 79:562–566, 1966.
550. Wolbach, S. B., and D. M. Hegsted. Perosis. Epiphyseal cartilage in choline and manganese deficiencies in the chick. A.M.A. Arch. Path. 56:437–453, 1953.
551. Wright, J. R., R. Levick, and H. J. Atkinson. Trace element distribution in virgin profiles representing four great soil groups. Soil Sci. Soc. Amer. Proc. 19:340–344, 1955.
552. Wurtman, R. J. Catecholamines and neurologic diseases. New Eng. J. Med. 282:45–46, 1970. (editorial)
553. Wurtman, R. J., C. M. Rose, S. Matthysse, J. Stephenson, and R. Baldessarini. L-dihydroxyphenylalanine: Effect on S-adenosylmethionine in brain. Science 169:395–397, 1970.
554. Wurts, T. C. Industrial sources of air pollution: Metallurgical, pp. 161–164. In Proceedings of National Conference on Air Pollution, Washington, D.C., November 18–20, 1958. Public Health Service Publication 654. Washington, D.C.: U.S. Government Printing Office, 1959.
555. Yarus, M. Recognition of nucleotide sequences. Ann. Rev. Biochem. 38:841–880, 1969.

Index

Absorption of manganese
 by plants, 51, 67-68, 72
 gastrointestinal, 79, 80, 82, 105, 134
 intoxication from, 79, 117, 135
 respiratory, 79, 105, 134
 through the skin, 79, 128, 135
Accumulation of manganese
 in animals, 80-81
 in man, 80-81
 in plants, 67, 68
Acidity of soil
 and ability to extract manganese, 68
 manganese concentration and, 51-53, 54, 70, 75, 133
 manganese deficiency and, 53, 54, 74
 manganese toxicity and, 54, 74
Adenosine triphosphate (ATP), 84
Age, at exposure to manganese, 104
Air
 inorganic manganese particles in, 141-44
 organic manganese particles in, 144
Air sampling, 16, 28-31, 141-42
 importance of station location, 29
 reliability of, 31
Air-dried soil, manganese solubility and, 6, 52

Airborne manganese
 and manganism, 105
 and pollution, 16-20
 factors influencing, 16
 from industrial operations, 16-28
 from mining operations, 26-27
 nonurban distribution, 34-35
 urban distribution, 32-33
Alcohol, and susceptibility to manganese, 104, 107
Alfalfa
 manganese concentration, 56, 63
 manganese tolerance, 72, 75
 manganese toxicity, 55, 57, 63, 65, 71, 74, 75
Algae, manganese in, 44
Alloys production, electrolytic manganese metal in, 13
Aluminothermic reduction, production of manganese metal by, 13
Aluminum production, manganese in, 12, 13
Ambient air manganese. *See* Airborne manganese
American Conference of Governmental Industrial Hygienists, 136
American Public Health Association, 145

Amino acids, effect on manganese toxicity, 65–66
Ammonium ion, effect on manganese uptake, 56
Analytic methods for manganese
  in air, 141–42
  in biologic materials, 142–48
  in blood, 145
  in hair, 147
  in urine, 144–45, 148
Animal behavior studies on manganese exposure, 121–23, 127–29, 136
Animal feed. *See* Forage
Animals, manganese tolerance, 78
Aplasia, manganese and, 88
Arginase activity, effect of manganese deficiency, 85, 91
Ataxia, from manganese deficiency, 87–88, 90, 116, 134
Atomic-absorption analysis, 134
  to analyze biologic samples, 147
  to trace manganese in air, 144
ATP. *See* Adenosine triphosphate
Automotive fuel, manganese in additives, 130–31
Automotive traffic, and manganese content of soil, 44
Auxiff, manganese toxicity and, 65

Bacteria
  to dissolve manganese, 37
  to leach manganese, 11
Banded manganese fertilizer, 54, 74
Basic oxygen furnace, 23, 24
Batteries. *See* Dry-cell batteries
Behavioral effects of manganese
  in animals, 121–23, 127–29, 136
  in man, 118–21, 134, 135
Bessemer process, 1, 24
Beverages, manganese in, 47
Biliary excretion of manganese, 80
Bioassay technique for analyzing biologic samples, 147–48
Biochemical abnormalities from manganese toxicity, 92–100, 135
Biochemical role of manganese, 83–87, 91, 134
Biogenic amines, effect on manganese metabolism, 98, 99
Biologic materials, analysis for manganese, 144–48

Biosphere, manganese in, 36–47
Birds, manganese tolerance, 78
Blast furnace, ferromanganese, 13. *See also* Ferromanganese blast furnace gas
  fume loading, 19
  location of, 19
  pollution from, 18–20, 49
Blood, analysis of manganese concentration, 145
Bog manganese ores, 11
Bone formation, manganese in, 77
Bones, effect of manganese deficiency, 85–86
Brain, manganese concentration, 105
Brain function, effect of manganese, 90, 91
Brain metabolism, effect of manganese poisoning, 93–100
Braunite, 5, 83

Calcium
  effect on manganese toxicity, 57, 72
  effect on manganese uptake, 55, 56, 64
Carbohydrate metabolism, 85, 89
Carbon dioxide fixation, manganese in, 85, 91
Carboxylation reactions, manganese and, 84
Cartilage, manganese-deficient, 86, 88
Cast iron production, manganese in, 12, 13
Catechol-*O*-methyltransferase (COMT), 95, 96
Catecholamines
  manganese toxicity and, 92–100
  urinary excretion of, 148
Cation-exchange capacity, 51, 54, 68, 75, 133
Cations
  effect on manganese toxicity, 57
  effect on manganese uptake, 56, 75
Cattle, manganese deficiency, 85–86, 89
Caudate nucleus, and manganese toxicity, 98
Central nervous system
  effect of manganese poisoning, 79, 82, 135
  importance of melanin to, 96
Ceramics, manganese in, 16

Index

Cereals
  manganese concentration, 59–61, 68, 69
  manganese deficiency, 68, 70
  manganese tolerance, 73
  manganese toxicity, 74
Chelatable manganese, 53, 56, 74, 82
Chemical forms of manganese, 14–16
Chlorosis, manganese-induced, 57, 64
Chondroitin sulfate, manganese for synthesis of, 77
Chronaxies
  effect of manganese, 118–19
  explanation of, 118
Chronic manganese toxicity, 92, 93, 99, 100
Cleaning systems
  for basic oxygen furnace gas, 24
  for blast furnace gas, 18, 25, 49
Clinical effects of manganese poisoning, 101–13
Coal-fired power plants, manganese emission from, 27
Color visibility threshold for manganese fumes, 20
COMT. See Catechol-O-methyltransferase
Condiments, manganese in, 47
Conduction–velocity studies, to determine effect of manganese poisoning, 119
Connective tissue, manganese and, 86, 88, 91
Consumption of manganese, 12
  alloys production, 12–13
  animal and poultry feed, 16
  ceramics, 16
  chemicals, 14–16
  dry-cell batteries, 12, 13–14
  fluxes, 16
  fuel additives, 16
  incandescent light bulbs, 16
  iron and steel production, 12–13
  metals production, 12–13
  paints, 16
  pharmaceuticals, 16
  U.S., 16–17
Copper, effect on manganese toxicity, 57
Cotton
  manganese concentration, 63, 68
  manganese toxicity, 55, 63, 71, 72, 73
Crops. See Plants
Cycling of manganese in the biosphere, 36–47
Cystathionase activity, manganese and, 148–49

Dairy products, manganese in, 45
Deep-sea sediment, manganese concentration, 4
Deoxidizer, manganese as, 12
Dermal contact and manganese intake, 79, 128, 135
Detoxification of manganese, 72–73
Diagnosis of manganese poisoning, 101–08, 112, 135–36
Diet
  as source of manganese, 77–78, 81
  results of manganese-deficient, 85–88, 91
Diethylenetriaminepentaacetic acid (DTPA), to diagnose manganese deficiency, 54
Digestive process, manganese in, 84
Dihydroxyphenylalanine (dopa), 97
Dissolved manganese, 37
  in river water, 42
Dissolved trace metals in water, 9, 11, 41
Distribution of manganese
  in body, 80–81
  in earth's crust, 3–5
  in minerals, 4–5
Dopa. See Dihydroxyphenylalanine
Dopamine
  depletion of, in Parkinsonism, 94, 97, 98, 99, 148
  metabolism of, 95–96
  normal brain content of, 93
Drainage basin, trace metals in, 38–39
Dry-cell batteries
  location of plants producing, 27
  manganese consumption, 12, 13–14, 48
Dry electrostatic cleaning system, for furnace gas, 24
DTPA. See Diethylenetriaminepentaacetic acid
Dust
  defined, 18, 49
  manganese, 102
    and manganic pneumonia, 110–11, 113
    and manganism, 102

Dust collector, on silicomanganese furnace, 20, 22
Dust emissions
 cleaning system for, 18, 24, 25, 49
 from manganese alloy production
  ferromanganese blast furnace, 18
  manganese concentration, 20, 21
  silicomanganese furnace, 20
 from pig-iron production, 25
 from steel production
  basic oxygen furnace, 23, 24, 25
  electric furnace, 23
  manganese concentration, 24, 25, 26
  open-hearth furnace, 23, 24, 25
  Thomas process, 24
 from storage of manganese ores, 26

Ear, inner, effect of manganese deficiency, 87–88, 90
Earth's crust, manganese distribution, 3–5
Ecosystem, manganese in, 3–5, 132
EDTA. *See* Ethylenediaminetetraacetic acid
EEG. *See* Electroencephalographic examination
Electric furnace
 ferromanganese production in, 13, 19–20
  manganese concentration in fumes from, 20, 22
 silicomanganese production in, 19
  manganese concentration in fumes from, 20, 22
 steel production in, 23, 25
  dust and fume emissions from, 23
Electroencephalographic (EEG) examination, to study effects of manganese intoxication, 120, 135, 136
Electrolytic manganese metal, 13–14, 48
 location of plants producing, 27
 production, 13
 uses for, 13
Electromyographic (EMG) examination, of effects of manganese poisoning, 119–20
Electrophoresis, manganese metabolism and, 56

Electrostatic precipitation, to collect manganese samples, 141, 142
Embryos, manganese deficiency in, 85
EMG. *See* Electromyographic examination
Emission spectrography, 7, 29, 45
Emission spectroscopy, 8
Emissions of manganese into the atmosphere. *See also* Manganese pollution
 factors influencing, 16–18, 31, 48–49
 from coal-fired power plants, 27
 from combustion of fuels, 17, 49
 from manganese alloy production, 18–22
 from steel and iron production, 23–26, 132
 severity of effects, 16–17
Environmental pollution with manganese, 126
Environmental Protection Agency, 129, 130
Enzymes
 effect of manganese, 55, 66, 83–85, 134
 in catecholamine metabolism, 95–96
 in manganese-deficient cartilage, 86–87
Epidemiology of manganese intoxication, 101–13
Ethyl Corporation, 129, 130
Ethylenediaminetetraacetic acid (EDTA), in treatment of manganese poisoning, 108–09, 145
Exchangeable manganese, in soil, 53, 54, 68
Excretion of manganese, 79–80, 82, 134
Exposure to manganese, 1–2
 from inhalation, 103–05, 113, 126–27
 range in safety levels, 114
 threshold limit for, 114, 115, 129, 139
 time period for, 104
Extraction of manganese, 54, 68, 69
Extrapyramidal diseases, 93–100

Fats, manganese in, 47
Federal Water Control Administration, 37
Ferroalloys, manganese in, 13
Ferromanganese production
 manganese in, 12, 13, 48
 total U.S., 20

Ferromanganese blast furnace gas, 18
  cleaning of, 18
  generation of, 20
  manganese concentration, 18
  variations in moisture content, 19
Fertility, effect of manganese, 89
Fertilizer
  banded manganese, 54
  manganese in, 15, 43, 44, 50, 51–52
  plant response to, 68, 70
Fish, manganese in, 46
Fly ash, manganese in, 27–28, 50
Food. *See also* Diet
  manganese concentration, 45–47, 77–78
Forage
  manganese in, 61–62, 66–67
  manganese supplement to, 15–16
Formalin, manganese solubility and, 52
Fresh water
  manganese concentration
    in solution, 9, 11
    in suspension, 9–11
    range of, 9
  problems in sampling and analyzing, 8
Fruit, manganese in, 46
Fuel additives, manganese in, 126–27, 128, 130–31, 133
Fuel combustion, manganese emission from, 17
Fumes
  collection of manganese samples from, 142
  defined, 18
  from manganese alloy production
    color visibility threshold for, 20
    from electric furnaces, 21
    from ferromanganese blast furnaces, 18
    size and shape of, 20, 21
  from steel and iron production, 23
    manganese concentration, 26
  particle size, 18
Furnace gas
  from manganese alloy production, 20
    amount generated, 20
    cleaning of, 18
    from blast furnaces, 18–20
    from electric furnaces, 19–20
  from steel and iron production
    cleaning of, 24, 25
  from basic oxygen furnaces, 23, 24
  from blast furnaces, 25
  from open-hearth furnaces, 23, 24

Gasoline additives, 126–27, 128, 130–31, 133
Gasoline combustion, manganese to improve, 126–27, 129–30, 132
Gasoline octane, manganese to improve, 126–27
Gastrointestinal system, absorption of manganese through, 79, 80, 82
Geographic distribution of manganese deposits, 5
Geographic variations in manganese concentration of soil, 6
Gluconeogenesis, manganese in, 88
Glucose tolerance, manganese and, 88–89
Glucose utilization, manganese in, 85
Goldschmidt process of aluminothermic reduction, 13
Grains, manganese in, 45
Grasses
  manganese concentration, 68, 69
  effect of washing, 44
  response to manganese fertilization, 70
Ground water, manganese concentration, 36–37
Growth retardation, manganese deficiency and, 90

Hair, manganese concentration, 147
Hemoglobin concentration, manganese and, 78–79
Herbs, manganese concentration, 68, 69
Horticultural crops, manganese concentration, 58–59
Hydroquinone production, manganese dioxide ore in, 14
Hypoplasia, manganese and, 88

Igneous rocks, manganese concentration, 3, 4
Impinger, to collect manganese samples, 141–42
Incandescent light bulbs, manganese in, 16
Industrial manganese. *See* Occupational hazard of manganese

Industrial plants, manganese emissions from, 16–17
Inhalation of manganese, 79, 83
  from environmental exposure, 126–27
  from industrial exposure, 103–04, 105, 113
  from manganese mining, 101
  manganic poisoning and, 110–11
Inorganic manganese particles, 141–44
Intelligence, manganese intoxication and, 117
Intestines, manganese concentration, 105
Intoxication
  from excessive manganese absorption, 79, 117, 135
  from mining, 78, 79, 101, 102
  from well water, 78, 109–10
  neurobehavioral effects of, 118–23
Iron
  compared with manganese, 3
  effect on manganese uptake, 56
Iron metabolism, manganese and, 64
Iron production, manganese in, 12, 13
Iron salts, effect on manganese toxicity, 57, 64
Iron uptake, 68
Iron-manganese ratios, for plant growth, 64
Irrigation water, manganese concentration, 43

Kidneys, manganese and, 105, 129

Lake and river water, manganese concentration, 11
L-dopa
  in treatment of manganism, 93, 99, 125, 135
  in treatment of parkinsonism, 93, 94, 96, 98–100, 135
Legumes
  manganese concentration, 69
  manganese tolerance, 71, 72
  manganese toxicity, 71
Levodopa. *See* L-dopa
Light, effect on manganese uptake and toxicity, 65
Lime, effect on manganese concentration, 51, 53, 55, 74
Liver, effect of manganese on, 89, 105, 129
Lung tissue sampling, 145
Lungs, manganese concentration, 105
Lupines, manganese concentration, 56, 68

Magnesium
  effect on manganese toxicity, 57
  effect on manganese uptake, 56
  manganese in production of, 12, 16
Manganese
  compared with iron, 3
  detoxified, 72–73
  dietary requirements, 77–78
  emissions. *See* Emissions of manganese
  functions, 55
  origin of word, 1
  valences, 3–4, 48
Manganese alloys
  consumption, 12–13
  pollution from production, 18–23
Manganese bronzes, 12
Manganese carbonates, 4
Manganese carbonyl compounds, 126–31
Manganese chemicals
  consumption, 14–16
  location of plants producing, 27
Manganese citrate, 56
Manganese complexes, 83–84
  enzymatic reaction of, 83
  magnetic moment of, 84
  stability of, 83
Manganese concentration
  geographic variations in, 5
  in animals, 78, 134
  in atmosphere
    nonurban, 29, 31, 34–35
    sulfuric acid generated from, 28
    urban, 29–33, 36, 132
  in fish, 44–45
  in foods, 45–47
  in lakes and river water, 9–11
  in man, 78, 80–81, 82, 134
  in mineral water, 11
  in plants, 51–56, 58–63, 132
    forage, 61–62, 66–67
    native Wisconsin plants, 67
    plant tops, 56, 66, 67, 68, 73

Index

in seawater, 7, 43-44, 48
  particulate versus dissolved, 7-8
in soil, 5-6, 51-55, 133
  regulation of, 74-75, 76
range of, 6
Manganese deficiency
  effect on arginase activity, 85, 91
  effect on brain functions, 90, 91
  effect on growth, 90
  effect on metabolism, 84-85, 88-89, 90, 91
  effect on skeletal system, 85-87, 116-17, 134
  in plants, 56-57
  in soil, 15, 53
  testing for, 53-54
Manganese deposits
  geographic distribution, 5
  in ocean floors, 6-7
Manganese dioxide
  production, 13-14
  use, 14-15
Manganese fertilizer, response to, 68, 70
Manganese metabolism, 55, 56, 148
  effect of biogenic amines, 98, 99
Manganese metals
  consumption, 13
  location of plants producing, 27
Manganese ores
  consumption, 12
  location of plants producing, 27
  manganese concentration, 26-27
  processing of, 12
Manganese oxide nodules, 6-7
Manganese oxides, 4, 48
  as absorbents, 28
  as catalysts, 28
  in fuel additives, 129, 130
  use of, 14-16, 48
Manganese poisoning. *See also* Manganism
  diagnosis of, 78, 79
  from industrial processes, 103-05, 107
  from mining manganese, 78, 79, 101-02
  from well water, 109-10, 113
  intoxication from, 78, 79, 101, 102
  neurologic symptoms of, 104-07, 109
  test for, 108
  treatment for, 108
Manganese pollution
  in Sauda, Norway, 20, 22-23
  of atmosphere, 18-23
  of seawater, 44-45
  of soils, 43
  of water, 37-43
Manganese salts, 5, 11
Manganese sulfate production, 14
Manganese supplementation, 87, 88, 89
Manganese tetroxide, 23, 24, 43
Manganese tolerance
  of mammals and birds, 78
  of plants, 71, 72, 73
Manganese toxicity. *See* Toxicity of manganese
Manganese uptake, by plants, 54, 68, 133
  factors influencing, 56, 70
  light and, 65
  temperature and, 65
Manganese-activated enzymes, 84
Manganese-zinc ferrites, 16
Manganic pneumonia, 79, 82, 110-11, 113, 135
Manganism. *See also* Manganese poisoning
  compared with parkinsonism, 93, 135
  from inhalation of manganese, 79, 83
  in manganese miners, 101-03
  neurologic symptoms, 106
  prognosis, 108
  psychomotor disturbances, 105
  treatment with L-dopa, 93, 99, 125, 135
Manganosilicosis, 102, 111
Manganous chloride, 14, 16
Manganous nitrate, 28
MAO. *see* Monoamine oxidase
Marine organisms, manganese in, 44-45, 50
Meat, manganese in, 46
Medical literature, on cases of manganism, 101-09
Melanin
  depletion of, in parkinsonism, 96-97, 99
  manganese concentration, 97, 98, 99-100, 135
Metabolism
  brain, 93-100
  carbohydrate, 85
  catecholamine, 95-96

Metabolism (continued)
  manganese, 55, 56, 148
  manganese deficiency and, 84–85, 88–89, 90, 91
  organic acids, 55
Metallic pollution of soil, 6
Metalloenzymes, manganese and, 84–85, 88
Metallurgic emissions, fume and dust content, 18
Metallurgic furnaces, manganese emissions, 18, 49
Metamorphic rocks, manganese concentration, 3, 4
Methylcyclopentadienyl manganese tricarbonyl (MMT), 16, 126
  antiknock capability, 130–31
  potential hazard from, 127
  production, 130
  toxicity, 128–29
Microbial metabolism, 11
Microorganisms, soil manganese and, 52–53
Mine waste, manganese contamination of soil, from, 43, 44, 50
Mine water, manganese concentration, 36–37
Mineral water, manganese concentration, 11
Mining of manganese, 12
  manganese poisoning from, 78, 79 101–02
    age distribution and, 104
    exposure period and, 104
Mitochondria, manganese in, 79, 80, 84–85, 89, 98, 134
MMT. *See* Methylcyclopentadienyl manganese tricarbonyl
Molybdenum, effect on manganese toxicity, 64–65
Monkeys, neurobehavioral effects of manganese intoxication in, 121–23
Monoamine oxidase (MAO), 95, 96, 97, 98
Mucopolysaccharide synthesis, 134–135

NASN. *See* National Air Surveillance Networks
National Air Surveillance Networks, (NASN), 29, 49, 136

National Center for Atmospheric Research, 36
National Research Council, 129, 130
Natural water, manganese concentration, 36
Nervous system. *See* Behavioral effects, Central nervous system, Neurobehavioral effects
Neurobehavioral effects of manganese intoxication, 118–24
Neurologic symptoms of manganese poisoning, 104–07
Neuromuscular electrostimulability, 127–29
Neutron activation analysis, 134
  of water samples, 148
  to analyze air samples of manganese, 143
  to analyze biologic samples, 146
Nickel toxicity, manganese and, 65
Nonurban stations, for air sampling program, 29, 34–35
Nutrition. *See* Diet
Nuts, manganese in, 46

Occupational hazard of manganese, 78, 79, 101–07
Oils, manganese in, 47
Open-hearth furnace, for steel production, 23, 24
Oral intake of manganese, 128. *See also* Diet
Organic manganese compounds, 144
Ornamental plants, manganese tolerance, 71
Oven-dried soil, 6
Oxidation of manganese, 64, 73, 83
Oxidative phosphorylation, 89–90

Paints, manganese in, 16
Pancreas, effect of manganese, 88–89
Parkinsonism
  compared with manganism, 93, 148
  L-dopa in treatment of, 93, 94, 96, 98, 99, 100, 135
  metabolic abnormalities in, 94–97
  relation between manganese toxicity and, 79, 92–100

Index

Particle size
  comparison of fume and dust, 18
  for manganese sample, 141
  in fume from electric ferroalloy furnace, 21
Periodate method
  to analyze air samples for manganese, 142–43
  to analyze biologic samples, 146
pH. *See* Acidity
Pharmaceuticals, manganese in, 16
Phenothiazine drugs, 97
Phosphorus
  effect on manganese solubility, 52
  effect on manganese toxicity, 65, 73
Photosynthesis, manganese in, 55
Pig-iron blast furance, 13, 25
Pig-iron production, 13, 26
Pigmentation, manganese and, 88, 93
Plants. *See also* Cereals, Forage, Horticultural crops, Ornamental plants
  and manganese, 51–76, 133
  manganese deficiency in, 53–54, 56–57
  manganese toxicity in, 55, 56, 57, 64–66
  tolerance to excess manganese, 70–72
Pneumonia, manganic, 79, 82, 110–11, 113, 135
Pneumonitis, 110–11, 113
Potassium permanganate
  manganese dioxide ores in production of, 14, 15
  manganic pneumonia and, 110–11
Poultry, manganese in, 46
Precipitation, manganese concentration, 36, 49
Pregnancy
  effect of manganese deficiency, 87
  manganese poisoning and, 125
Production of manganese, by country, 12, 14–15
Protein, manganese in, 81
Psychologic disturbances, from manganese absorption, 107, 135
Psychologic stress, and tolerance to manganese, 121
Psychomotor disturbances in manganism, 105, 135
Public health hazard, manganese as, 126–27, 129–30

Public water supplies, manganese concentration, 36, 37
  standards for, 41, 49
  violation of, 40, 43
Pyrolusite, 1, 4, 83
Pyruvate carboxylase, 84, 85, 88, 91

Radiomanganese, 79
Rainfall. *See* Precipitation
Rats, manganese deficiency, 86, 89, 90, 116
Reducible manganese, 53, 75
Reflexes
  effect of manganese, 118
  effect of noxious gases and dust, 117–18
Reports on urban–nonurban airborne manganese concentrations, 29–31
Reproduction, effect of manganese, 89, 91
Research suggestions, to assess biologic implications of manganese, 137–39
Reservoir water, manganese concentration, 37
Respiratory intake of manganese, 79, 82. *See also* Inhalation of manganese
Rhodochrosite, 4
Rhodonite, 5

S-adenosylmethionine, manganese and, 96
Samples, manganese
  analysis of, 142–44, 145–48
  ashing of, 145–46
  collection of, 141–42, 145
Seawater, manganese concentration, 7, 43–44
Sedimentary rocks, manganese concentration, 3, 4
Sensitivity to manganese. *See* Susceptibility
Sewage sludge, manganese contamination from, 44
Sex differences, susceptibility to manganese, 125
Shellfish, manganese concentration 44
Silicomanganese production
  in electric furnace, 13, 19
  dust collector on, 20, 22
  gas generated from, 20
Silicon, effect on manganese toxicity, 57
Size of particle. *See* Particle size

Skeletal deformities, from manganese deficiency, 85-86, 91, 116-17
Skin absorption of manganese, 79, 128, 135
Smoke inhibition, manganese for, 129-30, 132
Soil
  air-dried, 6, 52
  characteristics of, 51
  effect of waterlogging on manganese in, 73
  manganese concentration, 5, 43, 48, 132
  manganese contamination, 43
  manganese deficiency, 15, 53-54
  oven-dried, 6
  steaming of, 52
  trace-element analysis, 5-6
Soil extractants
  to predict manganese deficiency, 53-54
  to predict manganese toxicity, 54-55
Solubility of manganese, 51, 52, 54, 71
Soybeans
  manganese deficiency, 53
  manganese toxicity, 55
Spectrographic method
  to analyze biologic samples, 146
  to analyze human tissues, 45
  to analyze water, 7, 11, 37
  to determine manganese trace in air samples, 143
Spiegeleisen, 13, 48
Steaming of soil, manganese solubility and, 52, 53
Steel production
  manganese in, 1, 12, 13, 48
  U.S., 25
Sterility, effect of manganese, 89
Sulfur, manganese to nullify harmful effects, 12
Sulfur dioxide, 28
Sulfur trioxide, 28
Sulfuric acid, manganese in generation, 28
Surface water, manganese in, 36
Susceptibility to manganese
  and alcohol, 104, 107
  and development of manganism, 136
  and presence of infectious diseases, 104, 107-08
  by plants, 72-73, 75
  neuromuscular electrostimulability and, 127-28
  sex differences in, 125
Suspended manganese in water, 42
Suspended trace metals in water, 9, 11, 41
Swine, manganese deficiency, 85-86, 89, 116
Symptoms
  manganese deficiency
    in animals, 116
    in man, 134
    in plants, 56, 68
  manganese poisoning, 92, 104, 107, 108
  manganese toxicity, 57, 64, 74
  manganism, 103-08
Synthetic manganese dioxide, 13-14, 27

Temperature, effect on manganese uptake and toxicity, 65
Thomas process, in iron and steel production, 24
Thomas-slag pneumonia, 111-12
Threshold limit value (TLV), for manganese exposure, 129, 136
TLV. *See* Threshold limit value
Tobacco
  manganese concentration, 51-52
  tolerance to manganese, 71
Tolerance to manganese by plants, 70-72. *See also* Susceptibility
Toxicity of manganese, 28
  catecholamines and, 92-100
  effect on metabolism of brain, 93-100
  in animals, 93
  in environment, 127-28
  in man, 92-100
  in plants, 65
    concentration causing, 64-65, 74
    factors influencing, 55-56, 71-72
    methods of reducing, 57
    resistance to, 73
    symptoms, 57, 64, 74
  in soils, 52, 54-55
  role of particle size in, 115
  tests for, 124-25
Trace element
  analysis of soils for, 5-6

analysis of water for, 11, 37
  by drainage basin, 38-39
  manganese as, 78
Trace metals, spectrographic method for determination, 143
Transmethylation reactions, manganese and, 96
Treatment of manganism, 93, 99, 125, 135

Uptake. *See* Manganese uptake
Uranium ores, manganous dioxide ore for leaching, 14
Urban stations, for air sample program, 29-31, 32-33
Urinary excretion of catecholamines, 148
Urinary manganese concentration, 145
Urine sampling, 144-45, 148
U.S. Bureau of Mines, 11, 20, 23
U.S. Geological Survey, study of public water supplies, 36-37
U.S. Public Health Service, air sampling program, 28-31

Valences of manganese, 3-4, 83-84
Vegetables, manganese in, 46-47
Volcanic explosion, effect on airborne manganese, 29, 31

Washing, effect on manganese content of grass, 44

Water
  dissolved manganese in, 42
  manganese concentration, 48, 49, 132-33
    in groundwater and mine water, 36-37
    in irrigation water, 43
    in mineral water, 11
    in natural water, 36, 37
    in public water supply, 36, 37
    in reservoirs, 37
    in rivers and lakes, 9, 42
    in seawater, 7
    in wells, 43
    violation of standards for, 37, 40
  sampling and analysis of, 148
  suspended manganese in, 42
Water, trace metals concentration
  by drainage basin, 38-39
  dissolved, 41
  suspended, 41
Water system, infiltration of manganese into, 126
Waterlogging of soil, manganese tolerance and, 73
Well water
  manganese concentration, 43
  manganese poisoning from, 109-10, 113
Wet disintegrator cleaning system for furnace gas, 24

Zinc, effect on manganese toxicity, 57
Zinc production, manganese dioxide ore in, 14